DATE DUE			
FEB 2 4 1984			
DEC 0 9 1991			
FEB 1 7 1992			
AUG 3 1 1992			
APR 5 1993			
JUL 1 4 1993			
NOV 0 1 1993			
NOV 9 3 EXTD			
MAR 7 1994			
GAYLORD			PRINTED IN U.S.A.

FUGITIVE EMISSIONS AND CONTROLS

By
HOWARD E. HESKETH
FRANK L. CROSS, JR.

Copyright © 1983 by Ann Arbor Science Publishers
230 Collingwood, P.O. Box 1425, Ann Arbor, Michigan 48106

Library of Congress Catalog Card Number 82-72348
ISBN 0-250-40448-6

Manufactured in the United States of America
All Rights Reserved

Butterworths, Ltd., Borough Green
Sevenoaks, Kent TN15 8PH, England

The Environment And Energy Handbook Series

Howard E. Hesketh, PE, PhD
Editor-in-Chief

ANN ARBOR SCIENCE
THE BUTTERWORTH GROUP

PROPERTY OF U. S. GOVERNMENT

Copy No. 1

Preface—FUGITIVE EMISSIONS AND CONTROLS

In these times of stringent emission limitations, fugitive emissions are a significant area of concern. This handbook explains in sequence:

- how fugitive emissions are introduced into the atmosphere;
- the quantity and quality of typical emissions;
- control theories, practices and equipment;
- control techniques and combinations of techniques;
- calculation procedures for estimating emissions and control effectiveness;
- examples of controlled systems;
- emission measurement methods and equipment; and
- suggested record-keeping and reporting procedures.

In this book, we make no attempt to point out that fugitive emission control can be sound business practice, e.g, fugitive dust and gas losses can amount to thousands (and even millions) of dollars per year to individual operations, and reducing emissions can result in increased product yields, improved health and safer working conditions. Instead, we present the practical aspects of fugitive emission control with potential strengths and weaknesses along with examples of a few selected, but typical, industries. It is not possible in this short book to present examples for all industries and combinations of control systems. We suggest that readers use the presented theories and examples as guides to augment their own personal capabilities in determining possible optimum control strategies for specific situations.

Most of the discussions deal with particulate-type fugitive emissions. Gaseous fugitive emissions are also included in this handbook; however, an entire monograph will be required to adequately deal with these emissions, which are dominant in refineries, organic chemical plants and other operations.

Howard E. Hesketh
Frank L. Cross, Jr.

ACKNOWLEDGMENTS

The authors gratefully acknowledge the assistance provided by all the contributors of data, equipment photographs and drawings. We also thank Margaret (Mickey) F. Cross for manuscript preparation and Dr. Mohammad El-Shobokshy for proofreading.

Preface—THE ENVIRONMENT AND ENERGY HANDBOOK SERIES

This monograph series is unique. All the handbooks are written by specialists who are considered to be experts in their individual areas. As a result, the information presented can be considered as the "best available." The series is divided into a number of basic categories that cover a broad spectrum of subjects. Each book in itself is a complete work—within the narrow limits of the title. This relieves the authors from the need to present supplementary material that may not be within their specialized areas and enables the reader to concentrate in depth on the subject matter.

Discerning readers will acquire various monographs, as they are published, to build a complete literature resource base. For example, a reader interested in the control of industrial emissions using scrubbers would obtain that handbook. In addition, it could be useful also for this reader to obtain handbooks on costs, legal requirements, effects and other relevant subjects.

These monographs are both timely and of lasting value. As the need arises, a necessary area can be revised easily to ensure that the entire resource series will remain current.

The 15 subject categories comprising this series are:

1. Effects
2. Emissions and Regulations
3. Air Pollution Control
4. Water Pollution Control
5. Noise Pollution Control
6. Waste Pollution Control
7. Costs and Economics
8. Sampling and Monitoring
9. Conservation of Energy and Space (Including By-Product Utilization)
10. Operation and Maintenance
11. Advanced Technologies
12. Classical Energy Sources
13. Alternative Energy Sources
14. Education, Training and Opportunities
15. Supplementary Concerns

We sincerely hope that these handbooks become an integral part of your library and that they reflect our efforts to disseminate useful and needed information.

Editor-in-Chief

Howard E. Hesketh

Howard E. Hesketh, PE

Hesketh **Cross**

Howard E. Hesketh is Professor of Engineering, Air Pollution Control, at Southern Illinois University at Carbondale and author of the first air pollution textbook (published by Ann Arbor Science). He is also a consultant and advisor to the U.S. EPA, U.S. Department of Commerce, and other agencies, universities, organizations and industries.

Dr. Hesketh received his PhD in Chemical Engineering from the Pennsylvania State University, where he was a PHS Special Fellow. He is a Diplomate of the American Academy of Environmental Engineers and a registered Professional Engineer in Illinois and Pennsylvania. In addition to his practical and teaching experiences in air pollution, he received formal training in this area at the Pennsylvania State University Center for Environmental Studies.

Prior to his teaching, Dr. Hesketh was a Senior Chemical Engineer with Du Pont, where he was co-inventor of a fluidization and conditioning process for crystalline particles. He also worked with the Beryllium Corporation and Bell Laboratories for Western Electric.

He is a member of several professional societies, is on the Board of Directors and is currently a vice-president of the Air Pollution Control Association. He is an Associate Editor for ASME Transactions Journals, and has co-edited, authored or been contributing editor to several environmental texts. Dr. Hesketh is author of *Fine Particles in Gaseous Media* and *Air Pollution Control,* both published by Ann Arbor Science.

Frank L. Cross, Jr., with more than 25 years of diversified environmental engineering experience in industry and government, now heads his own consulting engineering and management firm, Cross/Tessitore & Associates, P.A., which specializes in air pollution control.

Before forming his own company, Mr. Cross was a Principal Consultant with Roy F. Weston, Inc., Consulting Engineers and Scientists, and Operations Field Services Administrator for the Department of Pollution Control in Florida. Mr. Cross has worked on environmental impact plans involving air and water pollution control, and noise abatement concepts. A professional trainer for many years, he was former Deputy Chief of the Training Program for EPA at Research Triangle Park.

Mr. Cross holds a ME in Air Pollution Control, a BS in Sanitary Engineering from the University of Florida, and a BS in Chemical Engineering from Northeastern University. He is a registered engineer in 10 states and is a Diplomate of the American Academy of Environmental Engineers.

CONTENTS

1. Fugitive Emissions 1

   1.1 Definition and Background 1
   1.2 Types 4
       1.2.1 Fugitive Dust Sources 4
       1.2.2 Nonparticulate Fugitive Emissions 5
   1.3 Dust Movement 5
   1.4 Wind Velocities 8
       1.4.1 Suspension Velocity—Regular Shapes 8
       1.4.2 Other Shapes 8
       1.4.3 Drift Extent 10
       1.4.4 Velocity Profile 10
   1.5 Sources 11
       1.5.1 Roads 13
       1.5.2 Tillage Operations 13
       1.5.3 Transfer and Conveyance 13
       1.5.4 Loading, Unloading and Sizing 13
       1.5.5 Aggregate Storage Piles 14
       1.5.6 Waste Sites 14
       1.5.7 Size Reduction and Construction 15
   1.6. Emissions 15
       1.6.1 Emission Data Reliability 15
       1.6.2 Uncontrolled Emission Factors 16
       1.6.3 Adjustment—Correction Parameters 31
       1.6.4 Special Adjustment of Factors 31

2. Control of Emissions 39

   2.1 Evaluating the Situation 39
       2.1.1 Introduction 39
       2.1.2 Source Identification 39
       2.1.3 Emission Assessment 40
       2.1.4 Required Limits 41
   2.2 Control Methods 41

      2.2.1 Control Options 41
      2.2.2 Control Techniques 43
      2.2.3 Preventative Procedure Controls 45
      2.2.4 Process Modification Controls 60
      2.2.5 Add-on Equipment Controls 60
      2.2.6 Example Arrangements 64
  2.3 Control by Industry 67
      2.3.1 Agricultural Operations 67
      2.3.2 Asphalt Concrete Plants 72
      2.3.3 Cement Manufacturing 75
      2.3.4 Foundries 75
      2.3.5 Grain Handling 75
      2.3.6 Lime Plants 75
      2.3.7 Mining Operations 77
      2.3.8 Ore Handling Facilities 78
      2.3.9 Coal-Fired Power Plants 87
      2.3.10 Nonferrous Smelting Operations 104
      2.3.11 Woodworking Operations 104

3. Measurement of Emissions 107

  3.1 Fugitive Dust Monitoring 107
      3.1.1 Monitoring Criteria 108
      3.1.2 Quasi-Stack Sampling 109
      3.1.3 Roof Monitor Sampling 109
      3.1.4 Upwind-Downwind Sampling 110
      3.1.5 Exposure Profiling 113
      3.1.6 Tracer Sampling 114
      3.1.7 Measurements 115
  3.2 Estimating 116
      3.2.1 Estimating Fugitive Emissions 116
      3.2.2 Estimating Control Efficiencies 122
      3.2.3 Models 124
  3.3 Reporting Fugitive Emissions 128

Appendix: Reference Method for the Determination of Suspended Particulates in the Atmosphere (High-Volume Method) 131

Nomenclature 141

Index 143

CHAPTER 1
FUGITIVE EMISSIONS

1.1 DEFINITION AND BACKGROUND

Fugitive emissions are considered to be those pollutants which are generated from open sources exposed to the air and are not discharged into the atmosphere in a confined flowstream. This definition includes gases as well as dusts. Most of the present concern is related to fugitive dust, so this monograph will concentrate on this area; however, gases will be discussed when appropriate. Some distinguish fugitive "emissions" as those from industry-related operations and fugitive "dust" as all other activities. Here these terms will be considered synonymous.

Air quality management concepts have emphasized point source emissions from stacks and emissions from transportation sources as the major sources of atmospheric pollutants. Within the past few years, with more emphasis on prevention of deterioration, control of fugitive emissions has become increasingly important. The inception of size-specific particulate standards would make fugitive emission control even more important. Particles less than 10 micrometers (μ) in diameter are inhalable and those less than about 2.5 μ are respirable [1,2].

Many areas of the United States do not meet the air quality goals of the State Implementation Plans (SIPs). Figures 1.1a and 1.1b show areas that do not meet the National Ambient Air Quality Standards (NAAQS) for total suspended particulates as of January 1, 1980 [3]. The primary U. S. NAAQS for total suspended particulates (TSP) as promulgated in the 1970 Clean Air Amendments are:

      75 μg/m³ annual geometric mean
    260 μg/m³ 24-hr maximum once a year

The TSP type of particles are basically those from 0 to 30 μ in diameter.

2 FUGITIVE EMISSIONS AND CONTROLS

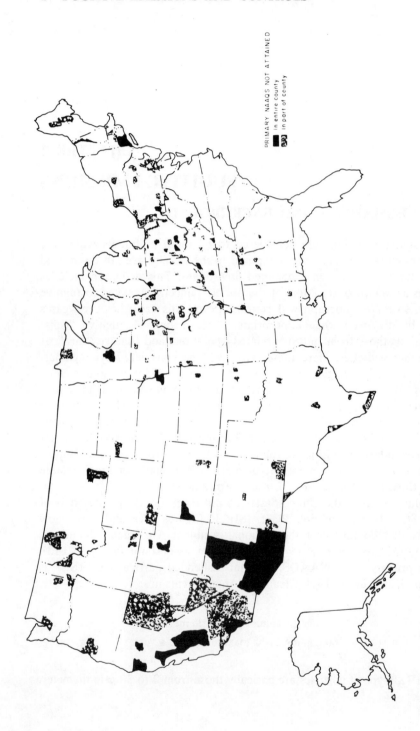

Figure 1.1a Nonattainment of primary NAAQS for total suspended particulates [3].

Figure 1.1b Nonattainment of secondary NAAQS for total suspended particulates [3].

4   FUGITIVE EMISSIONS AND CONTROLS

The 1977 amendments to the Clean Air Act established the Prevention of Significant Deterioration (PSD) increments, which for TSPs allow any existing actual annual geometric mean to increase by up to a maximum of:

$$5 \; \mu g/m^3 \text{ in Class I}$$
$$19 \; \mu g/m^3 \text{ in Class II}$$
$$37 \; \mu g/m^3 \text{ in Class III}$$

However, the NAAQS must not be exceeded. Class I areas are "pristine" areas (Indian reservations, national parks, etc.); Class II comprises moderate industrial growth areas; and Class III areas are major industrialized areas. Size-specific regulations for inhalable and respirable particulates have not been promulgated as of 1982; however, the proposed revised particulate standard suggests thoracic size particles $(0\text{-}10\mu)$ as the standard basis.

Although there are many legal issues related to NAAQS, PSD and fugitive emissions that are beyond the scope of this work, it is recognized that fugitive emissions contribute substantially to both TSPs and inhalable and respirable particulates.

1.2   TYPES

1.2.1   Fugitive Dust Sources

The bulk of fugitive emissions are fugitive dust particles. They are produced by mechanical disturbances of granular substances exposed to the air. For example, the disturbance can be caused by air moving at velocities in excess of about 12 mph or by pulverization and abrasion by some mechanical force. The basic sources of significant emissions are:

1. unpaved roads,
2. agricultural tilling operations,
3. aggregate storage piles,
4. mining, excavating and crushing operations,
5. industrial processing and transfer operations,
6. heavy construction operations, and
7. others—such as bare soil, unsealed landfills and evaporation of salt springs.

Some of the nonattainment areas of Figure 1.1 are industrialized. However, many are not, which indicates that fugitive dust of types 1, 2 and 7 are major sources. Rain suppresses these dusts. For comparison, note the typical precipitation isopleths as shown in Figure 1.2 [4]. On an average, the wind erosion rate is 8 tons/acre/yr in Colorado, New Mexico and Texas, and in the 10 Great Plains states it is about 5.3 tons/acre/yr.

1.2.2 Nonparticulate Fugitive Emissions

Gaseous emissions fall within fugitive emissions definition. They are quite significant in several aspects, yet they are usually placed in special categories and are not considered "fugitive emissions." Such special categories are odors and vapors. Odors from landfill, agricultural, industrial and other operations can be true gases (or vapors) or they may be adsorbed odorous gases on particles. Vapors, usually hydrocarbons, originate from both natural and man-made sources and can cause visibility problems and other adverse effects. Radioactive gases and hazardous materials are other special categories.

1.3 DUST MOVEMENT

Section 1.2.1 noted that fugitive dust can be picked up from the ground surface or other bed surfaces at wind speeds of about 12 mph or greater. Fugitive emissions are also generated from mechanical sources and can literally be thrown hundreds of feet into the air by both natural (e.g., volcanic) and anthropogenic (e.g., explosions, construction) sources. It is the particle size and density that affect the carrying capacity of the wind and hence the severity of the air quality problems. Ambient air quality basically relates to transport of the emissions past the specific source property line. Problems of emissions on the source site usually fall within the industrial hygiene area.

Once particles are set in motion due to either wind or mechanical forces, they can be transported by the wind. Depending on their shapes and densities, particles smaller than about 50 μ in diameter may remain suspended in the airstream for relatively long periods of time due to their buoyancy and the wind eddies. These suspended particles can travel large distances from their points of origin. The very small submi-

6 FUGITIVE EMISSIONS AND CONTROLS

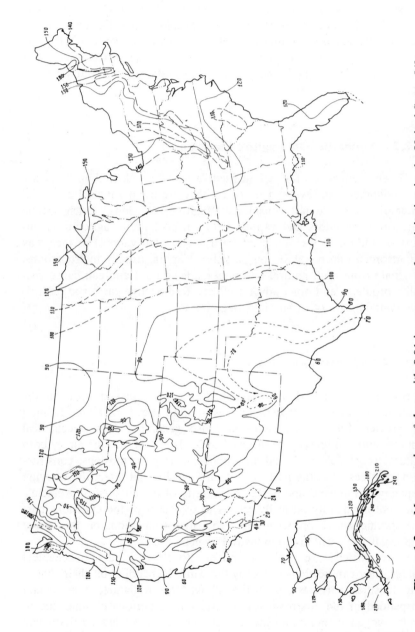

Figure 1.2 Mean number of days with 0.01 in. or more of precipitation in the United States [4].

cron particles act as gases and can remain essentially permanently suspended.

Depending on shape and density, particles larger than about 50 μ can become suspended, but they are not sustained in an airborne condition for long periods. These particles of fugitive dust are called "saltating" particles and after the initial wind or mechanical impulse settle via a "ballistic trajectory."

Constituting a third class of particles are those that move from the source area by a "surface creep" phenomenon: these large particles simply roll along the surface due to the wind. These particles have little adverse air quality effects and are not discussed further.

Dust particle sizes of interest range from 10-angstrom (Å) fog droplets to 100 μ dust particles (an angstrom is 10^{-10}m and a μ is 10^{-6}m, so 10 Å equal 0.001 μ). This is essentially a 10^4 size variation and a 10^{15} mass variation.

Baghold [5] shows that the velocity necessary to raise a particle into the airstream can be defined as the "threshold friction velocity"

$$U_{*_t} = k \left[\frac{d\, g\, \rho_p}{\rho_a} \right]^{1/2}$$

where d = particle diameter
g = acceleration due to gravity
ρ_p = particle density
ρ_a = air density
k = dimensionless parameter (a measure of mean fluid shear stress divided by the gravity stress on the particle)

Further, Hess [6] states that the "terminal velocity" of a particle settling from an airstream can be expressed as

$$U_f = \frac{2\rho_p g}{\mu_a} \left(\frac{d}{2}\right)^2 \left[\frac{24}{C_d\, Re}\right]$$

where μ_a = air viscosity
Re = Reynolds number
C_d = drag coefficient

The condition for suspension of a particle occurs when $U_f/U_{*_t} = 1$ This is when the "threshold friction velocity" equals the particle "ter-

8 FUGITIVE EMISSIONS AND CONTROLS

minal velocity." In essence, the "threshold friction velocity" is necessary for a particulate to become airborne, whereas the "terminal velocity" determines how long the particle will remain airborne.

For a neutral atmosphere, the frictional velocity and the free stream velocity are related by:

$$U_{*_t} = K_v U \left[\frac{1}{\ln(Z/Z_o)} \right]$$

where U = free stream velocity
K_v = von Karmen constant (~ 0.40)
Z_o = roughness parameter
Z = height of free stream velocity

Using the appropriate constants and the above relationships for U and U_{*_t}, and knowing the particle diameter and density, a value of U can be found to meet the suspension criteria of a given particle.

1.4 WIND VELOCITIES

1.4.1 Suspension Velocity—Regular Shapes

Using the relationships and data of Lee and Wilson [7] and others, a critical free stream velocity to suspend regular-shaped particles can be developed. This approximation is given as Figure 1.3 for normal atmosphere. These regular-shaped particles are assumed to be spherical, although small deviations would not be significant. For example, it requires a 2.1-ft/sec (1.4-mph) wind to suspend 50 μ particles (density 0.55 g/cm^3) and 2.8-ft/sec (1.9-mph) wind to suspend 50-μ dust (density 0.8 g/cm^3).

1.4.2 Other Shapes

If the particles in question have shapes that differ significantly from spherical, they will behave differently aerodynamically, and a correction must be applied. This is accomplished using:

FUGITIVE EMISSIONS 9

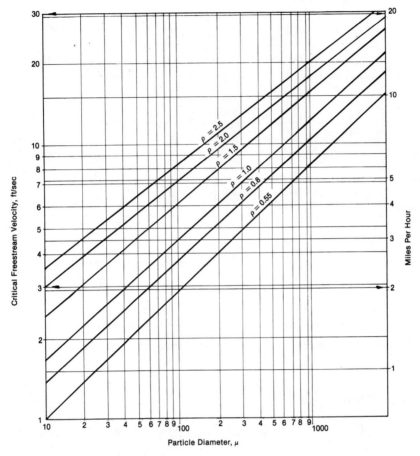

Figure 1.3 Critical freestream velocities required to suspend regular particles in neutral atmosphere.

$$d_A = 1.24 \frac{\sqrt[3]{V}}{\sqrt{\chi}}$$

where d_A = adjusted diameter for Figure 1.3
V = volume of particle
χ = shape factor

Values of χ are obtained from Figure 1.4 [8].

10 FUGITIVE EMISSIONS AND CONTROLS

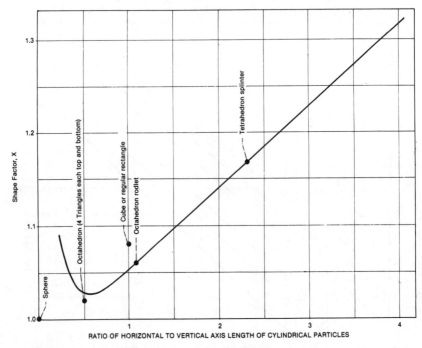

Figure 1.4 Ratio of horizontal to vertical axis length of cylindrical particles.

1.4.3 Drift Extent

Examples of road dust were developed by Cowherd et al. [9]. A particle density of 2.5 g/cm^3 was used. Particles traveling beyond 15 ft are influenced by atmospheric vertical velocity fluctuations. The calculated particle travel is given in Figure 1.5, but those traveling over 15 ft will have, on the average, drift distances greater than shown. Particles that travel about 85 ft will be small in size and remain suspended at normal wind speeds.

1.4.4 Velocity Profile

Wind speeds are assumed to exist at the measured or reference value except as noted in Section 1.5.5. These measurements are typically made at a height of 3 m (about 12 ft). The affects of dust suspension and drift reported in the previous section are based on typical conditions and measured wind speeds. It is important to remember that an actual wind

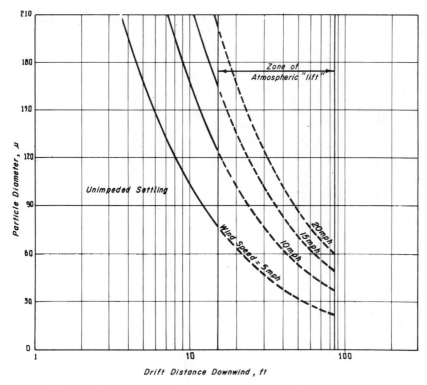

Figure 1.5 Calculated drift of typical unpaved road dust [9].

speed profile would look something like the one shown in Figure 1.6. This indicates that dust exposed to the wind at a 4-ft elevation actually sees a wind velocity 80% of the measured 3-m value, and dust less than 1 ft from the ground sees wind speeds less than 60% of the measured value. Figure 1.6 suggests that under these conditions and at normal wind speeds, most dusts would not become suspended. A mechanical assist or a wind gust would be needed to start the action.

1.5 SOURCES

Section 1.2.1 listed fugitive emission sources by type. This section attempts to categorize emissions by various operations, most (but not all) of which are industrially related. Emissions from natural and anthropogenic disasters are not included, although they can obviously be very significant.

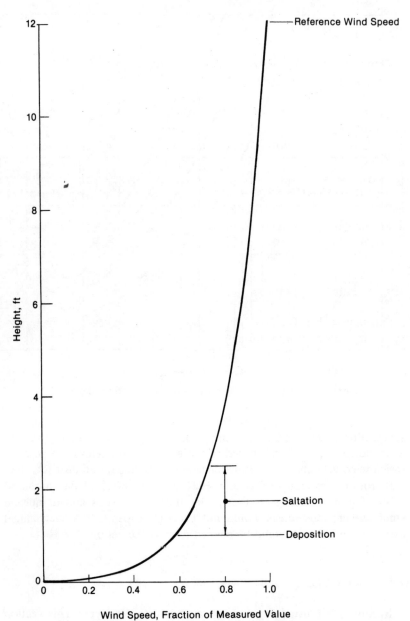

Figure 1.6 Typical wind speed profile.

1.5.1 Roads

Dust plumes are generated by vehicles traveling on unpaved (dirt and gravel) roads. The vehicle not only grinds the dust into finer particles, but also lifts and throws these particles into the air by the rotation of the wheels, the turbulent sheer air currents beneath the vehicle and the turbulent eddy air currents behind the vehicle. The amount of emissions depends on many factors, such as dust condition and type, vehicle size, speed, number of wheels, and type of tires and/or treads.

Paved roads are also sources of fugitive emissions. Figure 1.7 depicts several source and disposal mechanisms which in combination produce such emissions. Vehicle contributions can include tire and brake lining wear and exhaust. Shoulders of paved roads should also be considered as causes of significant emissions.

1.5.2 Tillage Operations

Substantial dust is raised when the surface of the earth is disturbed by implements traveling through or digging into the earth and by other devices used to break up the earth. Furthermore, dust continues to be emitted by subsequent actions. Dust amounts can vary greatly with the soil type and conditions and with the specific operation performed.

1.5.3 Transfer and Conveyance

These operations are present in nearly all industrial plants. The dominant emitters of fugitive dust are the belt conveyors, but also included are bucket elevators, screw conveyors, vibrating conveyors and screens. Open and elevated system losses and long, open devices are most significant sources of dust. Some transfer conveyors are over 12 mi long.

1.5.4 Loading, Unloading and Sizing

Receiving and disposing of dusty material is necessary in many industries and may actually be the initial and/or final transfer and conveying operation. This includes movement of bulk materials to or from rail cars, barges, trucks, ships and other transport systems. Fugitive dust is released as the material is mechanically agitated by the movement of the

14 FUGITIVE EMISSIONS AND CONTROLS

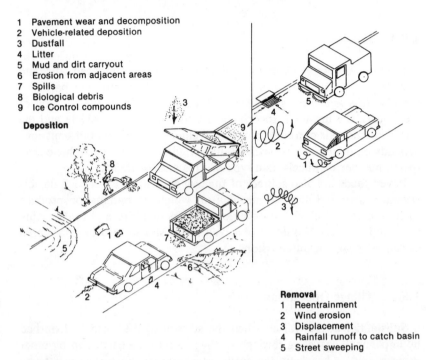

Figure 1.7 Paved road particulate deposition and removal mechanisms.

devices and by the turbulent air eddies created during the processes. Some loading procedures include a mechanical slinger that literally throws materials into the transport vessel to facilitate complete loading. Emissions from these are significant and difficult to control.

1.5.5 Aggregate Storage Piles

Storage piles that are open or only partially closed can be significant sources of fugitive dust. The materials stored are usually minerals such as coal, stone (sand and/or gravel), clay, gypsum, salts and ores. The materials in these piles are usually in constant rotation and therefore are stored for relatively short periods of time. Emissions are greatest during periods of material movement, high winds and dry periods.

1.5.6 Waste Sites

Process sludges and other wastes are frequently dumped in disposal sites for future reclamation or for relatively permanent placement. Some of these sites are concave holding areas, which are normally wet, but others are piles of the waste materials. Waste ponds include such materials as scrubbing system sludges and coal cleaning tailings. Piles can include overburden, process wastes, demolition material, bottom ash, fly ash and other air pollution control system discharges.

1.5.7 Size Reduction and Construction

Size reduction operations include mechanisms such as crushing and blasting. Construction is included because it is often preceded by demolition (size reduction). Significant amounts of force are exerted by these operations and, if they are not confined, fugitive dusts are released. Sometimes the dust is hurtled into the atmosphere to heights of hundreds of feet.

1.6 EMISSIONS

1.6.1 Emission Data Reliability

Section 1.6 summarizes emissions from various literature sources. All of these are obtained by some measurement or estimation technique and have reliabilities ranging from very poor to very good. Where more data are available and better tests have been conducted, the data are more reliable. The United States Environmental Protection Agency (U.S. EPA) has adopted a procedure for categorizing reliability ratings, and when appropriate and available we will use them. As given in AP-42 [10], emission data were scored from a low of 5 to a high of 40, and

Table 1.1 Reliability Rank of Emission Factors

Reliability Rank		Significance	
Numerical	Letter	Interpretation	Criteria
36–40	A	Excellent	Good field measurements of a large number of sources
26–35	B	Above Average	Good field measurements of a limited number of sources
16–25	C	Average	A minimum number of tests plus some engineering judgment
6–15	D	Below Average	A few tests plus engineering judgment
5	E	Poor	Engineering judgment based on experience with similar sources, perhaps some partial test data and physical observations

16 FUGITIVE EMISSIONS AND CONTROLS

then averaged and converted to a letter rank, as in Table 1.1. Criteria explanations given are the authors' interpretations.

The steps involved in estimating fugitive emissions at a plant consist of:

1. location of the source of emissions from a plant plot plan and flow diagram,
2. determination of the throughput composition and rate of each material at each point (e.g., lb/hr, tons/yr),
3. selection of appropriate emission factors and correction of these as needed for the specific conditions (see Sections 1.6.2 and 1.6.3), and
4. calculation and tabulation of the emissions for each source.

With fugitive dust emission estimates, certain special data may be required, such as the average wind speed in the area, silt content of the soil, amount of moisture and particle size of material. Further estimating procedures are presented in Section 3.2.

1.6.2 Uncontrolled Emission Factors

A definition of emission factor is "a typical value to indicate normal amounts of various and specific pollutants released from a given emission source when operated using specified operating procedures and/or devices and under specified conditions." Obviously, this is an attempt to quantify and qualify pollutants. Emission factors have been developed for many fugitive sources and those presented here are for uncontrolled sources. Reliability ranking for these is discussed in Section 1.6.1. These factors may be expressed on a basis of some quantity or rate and must be adjusted to relate to a specific system. Additionally, the factors may have to be corrected, as discussed in Section 1.6.3. Note in the following subsections that symbols are defined when used for the first time. If the meaning is not given with a particular equation, refer to the preceding sections or the Table of Nomenclature.

1.6.2.1 Roads

Fugitive dust from unpaved roads may average about 75 lb per vehicle mile of travel (VMT) [10] for a standard four-wheel automobile, depending on amount of silt, moisture and vehicle speed. Values for paved roads average about 6.1 g/VMT [11], with a range from 1–20 g/VMT. Obviously, specific conditions would greatly influence this factor.

The EPA emission factor, E, for four-wheel vehicles on unpaved roads [10] at 30–50 mph is:

$$E = (0.81)(s)\left(\frac{S}{30}\right)\left(\frac{365-w}{365}\right)$$

where E = lb of fugitive emissions/VMT
 s = silt content of road surface material, %
 S = average vehicle speed, mph
 w = mean annual number of days with 0.01 in. or more of rainfall (see Figure 1.8)

The silt content, s, is limited to 5 to 15% and is defined by the American Association of State Highway Officials as the particles <75 μ diameter (i.e., passing through a 200-mesh screen). The most significant dust fraction for total suspended fugitive dust values is the dust <30 μ in size. This is usually 60% of the total. Multiplying the previous equation by 60% and simplifying, we obtain:

$$E_{30} = (4.44 \times 10^{-5})(s)(S)(365-w)$$

where E_{30} = lb of <30 μ dust/VMT.

It is suggested [13] that emissions from trucks on haul roads would be greater than E_{30} by the prorating of the truck tire road surface to four wheel paved road car tire surface. Typical speeds on haul roads are 15 to 20 mph. An alternate technique for accounting for vehicle size is proposed by MRI [14] for unpaved roads. This is combined with the previous equation to give:

$$E_{30} = (1.48 \times 10^{-5})(s)(S)(365-w)W$$

where W = vehicle weight in tons

For paved roads this is

$$E_{30} = (3 \times 10^{-6})s\, L_1\, W$$

where L_1 = surface road dust loading on traveled portion of road in lb/mi

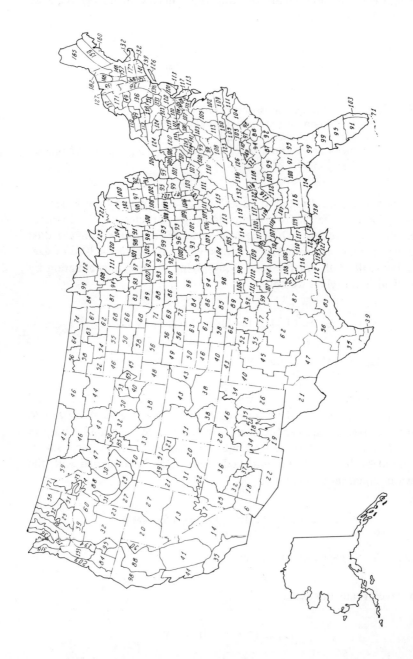

Figure 1.8 Thornthwaite's precipitation–evaporation index values for state climatic divisions [12].

FUGITIVE EMISSIONS 19

Example 1.1. Determine the amount of uncontrolled suspended emissions in tons/day from a fleet of two cars and ten trucks each making eight trips/day on a 7.5-mi haul road in northern Michigan at an average speed of 30 mph. The trucks weigh 12 tons. Assume road silt is 9%. Use w = 150 from Figure 1.2 [4,15].
Emissions from four wheel vehicles would be

$$(4.44 \times 10^{-5})(9)(30)(365 - 150) = 2.58 \text{ ton/VMT}$$

Emissions from trucks would be

$$(1.48 \times 10^{-5})(9)(30)(365 - 150)(12) = 10.31 \text{ ton/VMT}$$

Total emissions are

$$(2)(7.5 \times 2)(2.58) + (10)(7.5 \times 2)(10.31) = 1623.85 \text{ ton/day}$$

Emissions total $E_{30} \times$ VMT. VMT can be estimated by counting number of trips, one way, and multiplying by the distance; by use of fuel consumption and average vehicle mpg data; or by use of travel times and average mph data.

1.6.2.2 Tillage Operations

The advent of agricultural no-till farming techniques with chemical vegetation control should help reduce fugitive emissions as well as conserve energy. When tilling operations are used, emissions can be estimated by

$$E = \frac{500 \, s \, S \, T}{(PE)^2}$$

where E = annual emissions, lb/acre
s = silt content of soil, % (see section 1.6.2.1)
S = average machinery speed, mph (~5.5)
T = number of tillings per year
PE = Thornthwaite's precipitation–evaporation index (see Figure 1.8)

20 FUGITIVE EMISSIONS AND CONTROLS

This value of E excludes material settling within 20–30 ft of the tilling operation.

Heavy construction and excavation would be included as tilling operations. An approximate emission factor [10] for particles <30 μ is 1.2 tons/acre per month of activity when silt content is about 30% and PE equals 50. Emissions would vary directly with silt content and inversely with the square root of PE. MRI [14] suggests an experimentally determined emission rate, E_L, in lb/ton material moved, for front-end loader operations as:

$$E_L = (1.73 \times 10^{-3}) \frac{S\ U}{M^2 Y}$$

where U = mean wind speed, mph
 M = material moisture content, % by mass
 Y = loader bucket capacity, yd^3

Example 1.2. Estimate the annual uncontrolled emissions from a 500-acre farm in northern Michigan caused by three tilling operations per year. Assume soil silt content is 4% and use "average" tilling speed.

$$PE = 121 \text{ from Figure 1.8}$$

$$\text{Emissions} = \frac{(500)(4)(5.5)(3)}{(121)^2} = 2.3 \text{ lb/acre}$$

$$\text{Total emissions} = (2.3)(500) = 1150 \text{ lb/yr}$$

1.6.2.3 Transfer and Conveyance

Losses occur at the system feed and discharge points whether they are terminal or intermediate in location. If the system is exposed at any or all locations, losses occur due to spillage, windage and mechanical agitation. Because of the many system variations, possible emission factors can vary significantly and are not accurate on a general basis. PEDCo [16] summarize a number of emission factors for industrial sources. These are given in Table 1.2 along with other notes relative to material properties. These data are poor in that no data are recorded for: material characteristic size, material condition (wet or dry), length of transfer system, type of system, number of transfers occurring, type of feed and discharge and meteorological conditions. All of these must be considered to make these data more specific.

Table 1.2 Emission Factors for Transfer and Conveying Operations [15]

Material Name	Approx. Density, g/cm³ True	Approx. Density, g/cm³ Bulk	Emission Factor, lb/ton material transferred	Reliability Rank
Coal	1.25–1.45	0.67–0.91	0.04–0.96	E
Coke	1.75–2.00	0.75–1.1	0.023–0.13	E
Phosphate Rock (dry)		3.2	1.5	D
Sand		1.6–1.9	0.3	E
Grain		0.7–0.8	2.0–4.0	E
Iron Ore	4.9–5.2	3.5–5	2.0	E
Lead Ore	7.5	6.4	1.64–5.0	E

Example 1.3. Estimate the approximate uncontrolled emissions loss from a typical coal transfer conveyor during March if the conveyor coal average depth is 5 in., width is 16 in. and speed is 1.5 ft/sec. During March, winds and moisture would be higher. Therefore

$$\text{Assume } E = 0.96 \text{ lb/ton}$$
$$\text{Assume } \rho_B = 0.91 \text{ g/cm}^3$$
$$\text{Coal transferred} = \frac{(5)(16) \text{ ft}^2}{144} \left(\frac{1.5 \text{ ft}}{\text{sec}}\right) \frac{(62.4)(0.91) \text{ lb}}{\text{ft}^3}$$
$$\left(\frac{\text{ton}}{2000 \text{ lb}}\right) \left(\frac{3600 \text{ sec}}{\text{hr}}\right) = = 85.2 \text{ tons/hour}$$
$$\text{Emissions} = (0.96)(85.2) = 81.8 \text{ lb/hr}$$

1.6.2.4 Loading and Unloading

Often a similar (or the same) device is used for both loading and unloading. The amounts of emissions will vary depending on the type of device used, type of enclosure, material composition and moisture, material size, amount handled and meteorological conditions. The PEDCo summary of emission factors for loading/unloading [16] is given in Table 1.3. Note that the reliability factors show that these are relatively poor data.

Experimentally determined emission factors have been developed by MRI for loading operations [14]. Load-in stacker emissions, E_{IS}, in lb/ton material loaded are:

22 FUGITIVE EMISSIONS AND CONTROLS

Table 1.3 Emission Factors for Loading/Unloading Operations [16]

Material	Operation	Emission Factor, lb/ton material	Reliability Rank
Phosphate Rock (Dry)	Rail car and truck loading/unloading	1.5	C
Taconite Pellets	Rail car unloading in drive-through shed	0.03	E
Taconite Pellets	Ship loading with conveyor belts	0.02	E
Coal	Hopper car unloading; Barge loading	0.4	E
Grain	Loading/unloading:		
Grain	Rail car, drive-through shed	3–8	D
Grain	Truck, drive-through shed	2–8	D
Grain	Barge	3–8	D
Garbage	Truck unloading of large material	0.02	D

$$E_{IS} = 2.88 \times 10^{-4} \frac{s\,U}{M^2}$$

Load-out emissions, E_o, in lb/ton material loaded are:

$$E_o = 1.73 \times 10^{-3} \frac{s\,U\,K}{M^2 Y}$$

where K is an activity correction factor dependent on vehicular traffic in the area. Values of K are equal to 0.25 for iron ore pellets, coal and large stone and 1.0 for coke and screened stone.

Example 1.4. Estimate the uncontrolled emissions from transferring 85.2 tons/hr of medium-volatility coal from a pile to a conveyor belt using a front end loader with a "3-yd" bucket. Assume a 12-mph wind speed and a 7% moisture content.

Silt content = 2% (from Table 1.6)
K = 0.25
Y = 3 yd³

$$E_o = 1.73 \times 10^{-3} \frac{(2)(12)(0.25)}{7^2(3)} = 7.1 \times 10^{-5} \text{ lb/ton}$$

Total emissions = $(7.1 \times 10^{-5})(85.2) = 0.006$ lb/hr

1.6.2.5 Aggregate Storage Piles

Some of the emission factors given in the preceding sections are related to storage pile activities and should be included as appropriate. Additional factors are provided in this section which are specific to storage piles. The U.S. EPA estimates [10] fugitive emissions percentages from the four major storage pile operations, as shown in Table 1.4, for a typical stone pile with a 3-month storage cycle.

Table 1.4 Aggregate Storage Pile Emissions Expressed as Percentage Contribution by Operation [10]

Operation	Parameter	Approximate Activity Contribution, %
Vehicular Traffic	Rainfall, w	40
Wind Erosion	Climate, U	33
Load-out from Piles	PE	15
Load-in onto Piles	PE	12
Total		100

Cowherd, et al. in an early MRI study [17], gives the following equation for suspended dust emissions:

$$E_{30} = 3.3 \times 10^3 \frac{1}{(PE)^2}$$

where E_{30} is lb of particles <30 μ per ton placed in storage.

PEDCo data [16] suggest emission factors with a reliability ranking of D of 8.33 lb of particulates <30 μ per ton placed in storage (10.4 lb/acre of storage/day) for normal wind and pile activity of 5 days per week with 8–12 hr per day activity. A comparison of the PEDCo formulae [16] and MRI formulae [14] for storage pile fugitive dust emission factors is given in Table 1.5. In these formulae, symbols not previously defined are:

D_1 = duration of material storage, days
D_2 = number of dry days/yr
f = percentage of time wind exceeds 12 mph

24 FUGITIVE EMISSIONS AND CONTROLS

$K_{1,2,3}$ = vehicular activity factor; K is defined in the previous section
K_1 = 0.75 for coal, iron ore pellets and lumps
= 0.85 for coke
= 1.00 for slag and ore bedding
K_2 = 0.09 for iron ore pellets
= 0.4 for coke
= 0.5 for coal and ore bedding
= 1.00 for slag
K_3 = 0.75 for iron ore lumps and coal
= 0.85 to 1.00 for coke
= 1.00 for iron ore pellets, slag and ore bedding

Silt content of typical storage pile materials is given in Table 1.6. Note that treatment chemicals and additives to control emissions (including water) can cause some materials and/or material binders to break down, resulting in a significant *increase* in silt percentage. Excessive handling, wear and vehicular movement will also produce increases in silt.

Table 1.5   Storage Pile Fugitive Dust Emission Factor Formulae

	Emission factors, E_{30}, lb particulates <30 μ per ton of material transferred or stored as appropriate	
Operation	PEDCo Formulae [16][a]	MRI Formulae [14]
Vehicular Traffic Around Piles	$866.7 \dfrac{K_2 s}{(PE)^2}$	$8.51 \times 10^{-5} \, KsD_2$
Wind Erosion	$8.15 \dfrac{sD_1}{(PE)^2}$	$1.05 \times 10^{-7} \, sD_2 fD_1$
Load-out from Piles	$333.3 \dfrac{K_3 s}{(PE)^2}$	$1.73 \times 10^{-3} \dfrac{sUK}{M^2 Y}$ [c]
Load-in onto Piles	$266.7 \dfrac{K_1 s}{(PE)^2}$ [b]	$2.88 \times 10^{-4} \dfrac{sU}{M^2}$ [d]

[a] Preliminary formulae, reliability ranking of D.
[b] Thornthwaite's precipitation-evaporation index (PE) to be applied only to ore bedding piles; for all others, use PE = 100.
[c] Load-out assumed by front end loader.
[d] Load-in assumed by stacker.

Table 1.6   Silt Content, s, of Typical Storage Pile Materials

Material	Silt Content, %
Coke	1.0
Slag	1.5
Coal, medium-volatility	2.0
Coal, high-volatility	6.0
Iron Ore, lumps	9.0
Iron Ore, pellets	13.0
Ore Bedding	15.0

FUGITIVE EMISSIONS 25

Data from Carnes and Drehmel [18] present a procedure for estimating storage pile emissions due to wind erosion by including a function to account for pile orientation with respect to the wind. This emission factor in tons/yr is:

$$E = \frac{336(K_4 U_1)^3 K_5 \rho_B^2 A_1^{0.345}}{(PE)^2}$$

where U_1 = local wind velocity, m/sec
  ρ_B = storage pile bulk density, kg/m³
  A_1 = pile surface area, m²
  K_4 = correction factor for pile leading edge slope
    = 1.5 for θ of 30° (see Figure 1.9)
    = 1.0 for θ of 10° (see Figure 1.9)
  K_5 = sin ϕ (see Figure 1.10); this is the correction factor for projection of pile frontal area into plane perpendicular to wind direction.

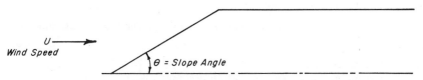

Figure 1.9 Effect of storage pile shape on fugitive emissions.

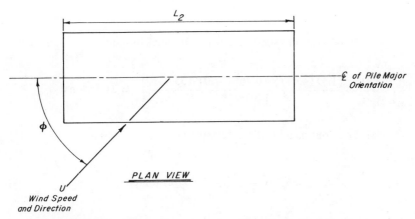

Figure 1.10 Effect of storage pile orientation to wind velocity.

26 FUGITIVE EMISSIONS AND CONTROLS

The Carnes and Drehmel correction factor K_5 is actually a velocity correction to account for pile orientation relative to wind direction. It is obvious that for an angle ϕ of $0°$, the equation would predict no emissions. Therefore, we recommend that ϕ of $11.5°$ be used for all angles less than $11.5°$ from the major axis.

Example 1.5. Estimate the annual uncontrolled emissions from a coke storage pile in Pennsylvania if the average pile size is 15 ft wide and 40 ft long. Assume the pile is oriented $50°$ to the mean wind direction and mean wind speed is 10 mph. Pile leading edge angle is $30°$. A total of 50,000 tons of coke is cycled through the pile annually:

$S = 1.0\%$ (from Table 1.6)
$K_2 = 0.4; K_3 \simeq 0.9; K_1 = 0.85$
$PE = 120$ (from Figure 1.8)
$\rho_B \simeq 1.0$ (from Table 1.2)
$K_4 = 1.5$
$K_5 = \sin 50° = 0.766$
$U_1 = (10 \text{ mph}) \left(\dfrac{0.447 \text{ m/sec}}{\text{mph}} \right) = 4.47 \text{ m/sec}$
$A_1 = (15)(40) \left(\dfrac{0.093 \text{ m}^2}{\text{ft}^2} \right)$
$\quad = 55.8 \text{ m}^2$ (neglecting the pile ends in this example)

Emissions from:
(a) Vehicular traffic using PEDCo formula is

$$886.7 \dfrac{(0.4)(1.0)}{(120)^2} = 0.025 \text{ lb/ton}$$

(b) Wind erosion using the Carnes-Drehmel formula is

$$\dfrac{(336)[(1.5)(4.47)]^3 (0.766)(1.0)^2(55.8)^{0.345}}{(120)^2} = 21.58 \text{ tons/yr}$$

(c) Load-out using PEDCo formula is

$$333.3 \dfrac{(0.9)(1.0)}{(120)^2} = 0.021 \text{ lb/ton}$$

(d) Load-in using PEDCo formula is

$$266.7 \frac{(0.85)(1.0)}{(120)^2} = 0.016 \text{ lb/ton}$$

Total emissions $= 21.58 + (0.025 + 0.021 + 0.016)\left(\frac{150{,}000}{2{,}000}\right)$
$= 26.17$ tons/yr

1.6.2.6 Waste Sites

Particulates. When applicable, emissions from wastes may be estimated using data from any or all of the sections on storage piles, tillage, size reduction, construction, loading and unloading, and transfer and conveyance. However, storage pile procedures are not applicable when the sites are ground level or below ground level. AP-42 [10] suggests a typical emission value using a construction criterion of 1.2 tons per acre of active construction per month. Prorate this for climatological factors using $\left(\frac{50}{PE}\right)^2$. The wind erosion emission factor equation in tons per acre per month is:

$$E_{30} = 0.011 \frac{e \, s \, f}{(PE)^2}$$

where $e =$ surface erodibility, tons/acre/yr

Gases. Gaseous emissions from waste sites can be significant and are therefore included here. Shen [19] gives the equation

$$E_i = D_i \, C_s \, A \, P_t^{1.33} \left(\frac{W_i}{L_D}\right)$$

where $E_i =$ emission rate of component i, g/sec
$D_i =$ diffusion coefficient of i, cm²/sec (see Table 1.7 and following discussion)
$C_s =$ saturation vapor concentration of i, g/cm³ (see following discussion)
$A =$ exposed area, cm²
$P_t =$ soil porosity, dimensionless (see following discussion)
$W_i =$ mass fraction of i, g/g
$L_D =$ effective soil cover depth, cm

Table 1.7 Diffusion Coefficients

Compound	Molecular Weight	Diffusivity in cm²/sec at °C	
		10	30
Vinyl Chloride	63	0.10094	0.11375
Benzene	78	0.08195	0.09234
Cyclohexane	84	0.07139	0.08045
Hexane	86	0.07021	0.07912
Toluene	92	0.07367	0.08301
Aniline	93	0.07157	0.08065
Xylene	106	0.06742	0.07597
PCB	189	0.04944	0.05571

Values of diffusivity for other materials can be estimated using values from Table 1.7 by

$$D_u = D_i \left(\frac{M}{M_u}\right)^{0.5}$$

where D_u = unknown diffusivity of gas in question
 M = molecular weight of gas for which diffusivity, D_i, is known, g/g-mol
 M_u = molecular weight of gas in question, g/g-mol

Values of the saturation vapor pressure, C_s, can be estimated using the ideal gas law by:

$$C_s = \frac{\rho M}{62{,}358\, T}$$

where ρ = vapor concentration of the gaseous chemical, mm Hg
 T = absolute temperature, °K

Soil porosity is estimated by the approximation

$$P_t = 1 - \frac{\rho_B}{\rho_P}$$

For example, with typical soil with a bulk density of 1.2 g/cm³ the porosity is

$$P_t = 1 - \frac{1.2}{2.65} = 0.55$$

When soil is compacted this may decrease to 0.35. Procedures for determining values of W_i may be found in publications such as those of the ASTM [20].

Example 1.6 (Particulates). Estimate the monthly uncontrollable emissions from a 1-acre waste site which receives fill once a week. The site is in Colorado. Assume silt content is 10%, erodibility is 45 tons/acre/yr and wind exceeds 12 mph 10% of the year.

$$PE = 38 \text{ (from Figure 1.8)}$$

Wind erosion is

$$E_{30} = 0.011 \frac{(45)(10)(10)}{(38)^2} = 3.1 \text{ tons/acre/mo}$$

Vehicular emissions equal

$$(1.2)\left(\frac{50}{38}\right)^2\left(\frac{1}{7}\right) = 0.30 \text{ tons/acre/mo}$$

Total emissions = 0.50 tons/mo.

Example 1.7 (Gases). Estimate the uncontrolled emissions of benzene in g/sec from a 0.1-acre waste site containing 1000 ppm benzene by mass at 10°C. The vapor pressure of benzene at this temperature is 44 mm Hg from handbook physical property data. Three inches of soil with a porosity of 0.5 covers the area.

$W_i = 0.001$
$D_i = 0.08195 \text{ cm}^2/\text{sec}$ (from Table 1.7)
$M = 78 \text{ g/g-mol}$
$T = 273 + 10 = 283$
$P_t = 0.5$

$$C_s = \frac{(44)(78)}{62{,}358(283)} = 1.94 \times 10^{-4} \text{ g/cm}^3$$

$$A = (4.05 \times 10^7 \text{ cm}^2/\text{acre})(0.1) = 4.05 \times 10^6 \text{ cm}^2$$

$$L_D = (3)(2.54) = 7.6 \text{ cm}$$

$$E_i = (0.08195)(1.94 \times 10^{-4})(4.05 \times 10^6)(0.5)^{1.33}\left(\frac{0.001}{7.6}\right)$$

$$= 3.37 \times 10^{-3} \text{ g/sec or } 3{,}370 \text{ }\mu\text{g/sec}.$$

1.6.2.7 Size Reduction and Construction

The previous section discusses the construction activity emission factor of 1.2 tons/acre of active construction per month and the correction factor of $\left(\frac{50}{\text{PE}}\right)^2$ for climatic conditions. Many operations that produce particle size reductions have no reliable emission factors. Construction aggregate crushing operations, however, have been sampled, and emission factors have been determined. Variations from the reported values will occur due to wind speeds and with type of specific operation (e.g., wet or dry). The National Crushed Stone Association [21] data are used for the values given in Table 1.8. The 10-μ values are measured by trace analysis or by source assessment, and the 30-μ values are estimated using size distribution data from Calvert [22] and NCSA.

Table 1.8 Particulate Emission Factors for Stone Crushing Operations

Operation	Emissions in lb/ton of particulate		Emissions from AP-42 [10], lb/ton
	<10 μ	<30 μ	
Primary Crushing			0.10
Granite	0.01	0.16	
Limestone	0.0003	0.0009	
Sand & Gravel	0.0008	0.0023	
Traprock	0.0026–0.0008	0.0075–0.0023	
Secondary Crushing			0.6
Granite	0.0006–0.02	0.10–0.33	
Limestone	0.0001–0.064	0.001–0.724	
Sand & Gravel	0.002	0.024	
Traprock	0.001	0.012	
Tertiary Crushing			3.6
Granite	0.08	1.30	
Limestone	0.006	0.023	
Traprock	0.002–0.008	0.024–0.0970	
Finer Crushing			4.5
Granite	0.008	0.130	
Traprock	0.0040–0.0014	0.017–0.048	

1.6.3 Adjustment—Correction Parameters

1.6.3.1 Obtaining the Factor

Many correction factors are given in Section 1.6.3 to account for some of the variations in conditions and operations. Some typical values for these factors are also provided and can be used as appropriate. Often, it is advisable to obtain specific values for the system in question. Laboratory and/or field analyses can provide quick information. Locally available data should also be used as much as possible, depending on reliability and suitability.
Examples of these are:

Silt content—Measure the amount of loose, dry *surface* dust that passes through a 200-mesh screen.
Moisture content—Heat a sample of material with a heat lamp 4–6 in. away to constant weight.
Storage duration and process rates—Use estimates from plant personnel and process data.
Climatic data—Airport, plant and/or local weather data could all be available. If not, check U.S. Weather Bureau data for the surrounding stations' data. All wind speeds should be low-level (3 m or less) and at an elevation similar to the emission site in question.
Activity factor—This is a judgment estimate and as such must be considered as a "special adjustment," as described in the following section.
Particle size—Emission of material <30 μ can vary depending on particle size and distribution. For example, average unpaved road dust contains 35% by mass of 2- to 30-μ material and 25% by mass of material <2 μ, for a total of 60% material <30 μ. If the road were new with fresh crushed stone, the value could be much higher. Also, if the stone were very hard, the amount of material <30 μ could be lower.

1.6.4 Special Adjustment of Factors

Emissions based on emission factors will be inexact because of the reasons given above as well as others not noted. Therefore, when unrealistic numbers are encountered, it may be necessary to make engineering adjustments in these numbers. An example of this occurs when during calibration of a dispersion model, the predicted ground level concentrations (GLCs) do not correlate with measured GLCs. One of the factors

32 FUGITIVE EMISSIONS AND CONTROLS

directly affecting this discrepancy may be the emission inventory. In cases where this is suspected, investigators should determine particle size, settling velocities, and reassess:

1. how much of the emission is actually airborne,
2. whether any of the correction factors need to be adjusted, and
3. whether the activity correction is appropriate.

Case History 1.1

Cowherd, et al. [17] report a study of estimated emissions vs measured actual emissions of dust <30 μ downwind from an unpaved road (Table 1.9). As expected, when silt content is measured, estimated and measured results are quite close. When silt content is estimated, the results are more often significantly in error and may be quite inaccurate even though all other corrections are exact. Notice the very high silt content of the last test and the good estimated value.

Case History 1.2

The following case history studies indicate how fugitive emission estimates for taconite storage piles were revised based on field and laboratory investigations/tests and a literature evaluation. In studying the fugitive emissions from a taconite ore dumping, storage and loading terminal, the EPA Manual (EPA-450/3-77-010, "Technical Guidance for Control of Industrial Process Fugitive Particulate Emissions") was the primary basis for estimating quantitative fugitive emission levels for all material handling and storage processes at the facility. Specifically, computer modeling of ground level particulate impacts outside of the facility are based on the above quantitative emissions becoming airborne and suspended, and later redeposited at a remote location. Further investiga-

Table 1.9 Comparison of Estimated and Actual Unpaved Road Emissions [17]

Vehicle Speed, mph	Silt, %[a]	Emission Factor, lb/Vehicle Mile Traveled		Error, %
		Estimated	Actual	
30	12 m	9.7	10.0	−3
30	13 m	10.5	10.3	2
40	13 m	14.0	13.9	1
30	20 e	16.2	16.3	−1
40	5 e	5.4	6.0	−10
30	68 m	55.1	55.9	−1

[a] m = measured, e = estimated.

tion reveals that all airborne particulate does not become suspended and carried to a remote location.

The purpose of this study was to determine the potential fugitive particulate size distribution for a specific facility, and the effect of this distribution on the emission rates used in the previous air impact modeling studies. In general, the study consisted of the following:

1. representative acquisition at the terminal,
2. laboratory analysis to determine particle size distribution,
3. review of particle size distribution data and comparison with published data for potential suspended particulate, and
4. adjustment of original emission factors based on size distribution and "particle suspension" criteria.

Sample Acquisition Procedure. Twelve taconite ore/dust samples were collected. The following table defines the location and nature of each. In general, these consisted of ten taconite pellet samples and two dust samples (Nos. 5 and 7) as shown in Table 1.10. Three types of ore are shown as types X, Y and Z.

Duplicate storage pile samples were collected for X and Y pellets. Z pellet samples were obtained from railroad cars prior to dumping,

Table 1.10 Results of Grain Size Distribution Analysis for Taconite Ore Terminal

Field Sample No.	Type of Ore	Sampling Location	Sample Preparation	% by Weight $<100\,\mu$	% by Weight $<50\,\mu$	% by Weight $<30\,\mu$
1	X	RR Car	Pellets Ground	78	54	45
2	X	RR Car	Pellets Ground	84	60	52
3	Y	Pile (Top)	Pellets Ground	55	34	24
4	Y	Pile (Side)	Pellets Ground	52	31	20
5		Below Conveyor No. 8	Dust—No Preparation	60	28	18
6	X	Stacker Discharge	Pellets Ground	72	50	42
7	X	Pile Near Stacker	Dust—No Preparation	24	12	8
8		On Stacker and Conveyor	Pellets Ground	65	40	30
9	Z	Pile	Pellets Ground	75	60	45
10	Z	Pile	Pellets Ground	68	46	35
11	X	Transfer Point Conveyor 2 to 3	Pellets Ground	70	50	41
12	X	Transfer Point Conveyor 2 to 3	Pellets Ground	73	55	40

from a transfer point (Conveyor 2 to Conveyor 3) and from the storage pile beneath the stacker.

Procedure consisted of shoveling either the pellets and/or dust into dust-free plastic containers, sealing the containers, marking each container for identification, and delivering the containers to the testing company for analysis. Each sealed container held greater than 50 lb of taconite pellets and/or dust.

Laboratory Analysis Procedure. The laboratory analyses consisted of the following steps:

1. Each sample required approximately 5000 g of dust for accurate sieve analysis. Therefore, in the case of the ten containing mainly pellets (where the dust was insufficient), the pellets were placed in a Los Angeles Abrasion Machine and rotated for 20 min with five spheres to break down the pellets and generate dust. This procedure was not necessary in the cases of Samples 5 and 7, which contained sufficient dust.
2. The taconite ore dust was then sieved on #40 sieve.
3. All material passing the sieve was then further sized using a hydrometer.
4. The hydrometer results were then plotted on a standard grain size analysis sheet.

It should be noted that the hydrometer computations are based on an average specific gravity of solids determined to be 4.7 in general accordance with American Society of Testing Methods (ASTM) D 422. In actuality, the dust consists of iron ore and bentonite particles, which individually have different specific gravities.

Laboratory Results. Figure 1.11 is indicative of one of the typical laboratory size analyses. The following listing summarizes the results of the particle size analyses.

A review of the size distribution figures and the summary listing show the following:

1. The two dust samples (Nos. 5 and 7) show a distribution skewed to large particle distribution when compared to dust from the ground pellets. This is probably due to the fact that the existing dust piles consist of the large particles which have resisted becoming airborne and suspended. The smaller particles (less than 50 μ) have already been removed due to wind erosion.
2. If the company Y pellet samples are eliminated, the remaining pellet samples have the following statistical properties:

FUGITIVE EMISSIONS 35

No. of Samples	Mean Value, % by wt <100 μ	Mean Value, % by wt <50 μ	Mean Value, % by wt <25 μ
8	73.1 (range, 65–84)	51.8 (range, 40–60)	41.2 (range, 30–52)

3. If the company Y pellet samples are included, the statistical properties for all pellets collected would be as follows:

No. of Samples	Mean Value, % by wt <100 μ	Mean Value, % by wt <50 μ	Mean Value, % by wt <25 μ
10	69.2 (range, 52–84)	47.9 (range, 31–60)	37.3 (range, 20–52)

Potential Emissions Based on Size Distribution Results. Assume that emissions consist of particles below 100 μ. The literature shows analytically that airborne fugitive emissions do not become suspended unless the terminal velocity of the particle (U_f) equals the threshold friction velocity (U_{*_t}), or (U_f/U_{*_t}) = 1. This is shown in Figure 1.12. These curves show the analytical break point for suspension and deposition for

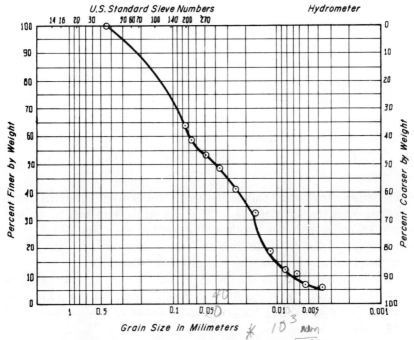

Figure 1.11 Ore dust sample from railroad car.

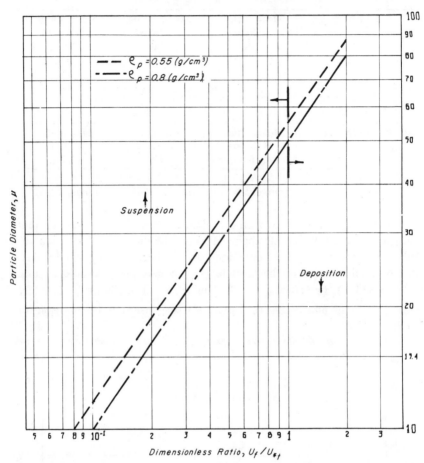

Figure 1.12 Particle diameter vs. ratio of terminal velocity to threshold velocity.

two particle densities. Analytically, they show that for the given densities, particles ranging from 48 to 52 μ are on the threshold of suspension. Therefore, particles less than 50 μ will become suspended.

In reality, the effect of moisture and particle irregular shape may cause the threshold for suspension to occur at lower diameters.

If it is assumed that the original EPA fugitive emissions are based on particles less than 100 μ and a suspension level of 50 μ and 30 μ is assumed, then the following correction factor can be applied to the original emissions:

CORRECTION FACTOR

Samples	50-μ Suspension Threshold	30-μ Suspension Threshold
Pellets X and Z	0.71	0.56
Pellet Y	0.69	0.54

Conclusions. Based on the above sampling data, the suspension threshold criteria and the fugitive data in normally used emission estimates may be corrected for both 50-μ (theoretical suspension threshold) and 30-μ (realistic suspension threshold) particles.

REFERENCES

1. Amdur, M. O., and M. Corn, "The Irritant Potency of Zinc Ammonium Sulfate on Different Size Particles," *Am. Ind. Hyg. Assoc. J.,* Vol. 24, July-August 1963, pp. 326-333.
2. "The Technical Basis for a Size Specific Particulate Standard, Parts I and II," Specialty Conference, Chatten Cowherd, Jr., Chairman. Edited by the Air Pollution Control Association, APCA, March and April 1980.
3. Bunicore, A. J., "Air Pollution Control," *Chem. Eng.,* Vol. 87, No. 13, June 30, 1980, pp. 81-101.
4. Jutze, G. A., K. Axetell, Jr. and W. Parker, "Investigation of Fugitive Dust—Sources Emissions and Controls," PEDCo Environmental Specialists, Inc., Cincinnati, OH. U.S. EPA Publication No. EPA-450/3-74-036a, June 1974.
5. Bagnold, R. A., "The Physics of Blown Sand and Desert Dunes," (London: Methuen, 1941).
6. Hess, S. L., "Dust on Venus," *J. Atmos. Sci.,* 32:1076, 1975.
7. Lee, Y., and J. W. Wilson, "Windblown Fugitive Particle Emissions from Storage Piles," paper no. 79-11.2, 72nd Annual APCA meeting, Cincinnati, OH June 1979.
8. Hesketh, H. E., "Air Pollution Control," (Ann Arbor, MI: Ann Arbor Science Publishers, Inc. 1981).
9. Cowherd, C., Jr., T. H. Southerland and C. O. Mann, "Development of Emission Factors for Fugitive Dust Sources," proceedings 67th Annual APCA, Denver CO, June 1974.
10. "Compilation of Air Pollutant Emission Factors," U.S. EPA #AP-42 with supplements, February 1976.
11. Midwest Research Institute, "Quantification of Dust Entrainment from Paved Roadways," U.S. EPA Contract No. 68-02-1403, Task Order No. 7, March 31, 1976.
12. "Climatic Atlas of the U.S.," U. S. Department of Commerce, Environmental Sciences Services Administration, Environmental Data Service, Washington, DC, June 1968.

13. PEDCo Environmental Specialists, Inc., "Evaluation of Fugitive Dust Emissions from Mining," Task 1 Report; U.S. EPA, Contract No. 68-02-1321, Task No. 36, April 1976.
14. Midwest Research Institute, "A Study of Fugitive Emissions from Metallurgical Processes," U.S. EPA Contract No. 68-02-2120, November 4, 1977.
15. Thornthwaite, C. W. "Climates of North America According to a New Classification," *Geog. Rev.* 21, pp. 633-655, 1931.
16. PEDCo Environmental, Inc., "Technical Guidance for Control of Industrial Process Fugitive Particulate Emissions," U.S. EPA No. EPA-450/3-77-010, March 1977.
17. Cowherd, C., Jr., K. Axetell, Jr., C. M. Guenther and G. A. Jutze, "Development of Emission Factors for Fugitive Dust Sources," Midwest Research Institute, U.S. EPA No. EPA-450/3-74 037, June 1974.
18. Carnes, D., and D. C. Drehmel, "The Control of Fugitive Emissions Using Windscreens," The third US EPA Symposium on The Transfer and Utilization of Particulate Control Technology, Orlando, March 1980.
19. Shen, T. T., "Estimating Hazardous Air Emissions from Disposal Sites," *Poll. Eng.*, August 1981, pp. 31-34.
20. Conway, R. A., and B. C. Malloy, Eds., "Hazardous Solid Waste Testing: First Conference," ASTM/STP 760, American Society of Testing Methods, Philadelphia, PA, December 1981.
21. "An Investigation of Particulate Emissions from Construction Aggregate Crushing Operations and Related New Source Performance Standards," National Crushed Stone Association, Washington, D.C., December 1979.
22. Calvert, S., et al., "Scrubber Handbook," U.S. EPA Contract No. CPA-70-95, July 1972.

CHAPTER 2
CONTROL OF EMISSIONS

2.1 EVALUATING THE SITUATION

2.1.1 Introduction

Fugitive emissions do exist, and there are usually several techniques that can be used to reduce the amount of material emitted. A number of the control techniques utilize some of the same principles involved in the control of confined source emissions, while many others are applicable only to fugitive sources.

In addition to the legal requirements, some reasons fugitive emissions should be controlled are:

> *Large particles*—Particles <100 μ may be windblown and can cause nuisance problems.
> —Particles <30 μ will be suspended and can cause nuisance problems.
> —Particles <15 μ are "inhalable" [1], and those with a density of 2.5 g/cm or less can be transported over long distances.
> *Fine particles*—Particles <2.5 μ are respirable [1] and cause adverse health effects. They are suspended and can be transported long distances.
> —Particles <1 μ are responsible for most atmospheric visibility problems, although particles of any size may cause local visibility problems.

2.1.2 Source Identification

Actual fugitive dust emissions and sources vary for reasons given in Chapter 1, but a typical industrial fugitive emission may have source contributions of <30 μ dust of:

unpaved roads and carry-on dirt to paved roads	70%
wind erosion of open areas	5%
storage piles losses	25%

Industrial fugitive emissions frequently contribute more than 50% of the total suspended and inhalable particulate emissions [2]. In addition, these particulates frequently contain toxic or hazardous substances.

Each industry has a specific group of fugitive emission sources with different shapes, sizes, and emission rates. However, storage piles, conveyors, transfer points, and haul roads are common to many industries. The following listing gives the approximate ranking of operations as fugitive emissions sources in order of decreasing emission rates for the mining industry [3]:

1. overburden removal
2. haul roads (unpaved)
3. land reclamation
4. storage piles (active and inactive)
5. loading
6. transfer and conveying
7. unloading
8. blasting
9. crushing
10. waste disposal
11. coal cleaning

Other possible operations that could be added to this listing of fugitive emission sources are:

- exposed areas (parking lots, unused land, agricultural fields, recreation areas);
- paved roads (spilled and/or tracked out dirt); and
- construction sites.

2.1.3 Emission Assessment

In this chapter we will be dealing not only with specific operations, but also with the combinations of operations that make up a specific facility and the emissions associated with that total facility. Basically, the problem and solution can be divided into three stages. The first stage consists of gathering data relative to the facility. This consists of applying emission factors to each source using best available meteorological data and onsite information relative to the operations, the properties of the materials and the topographical parameters of the site. Using these data, the fugitive emission impact of the facility can be estimated.

If the Stage 1 evaluation indicates that a problem exists or if complaints make it necessary to proceed further, then Stage 2 action should be incorporated. This consists of making additional field and laboratory measurements and observations to correct and refine the emission data. It may be useful to model problem area sources and/or set up monitoring systems as described in Chapter 3 to test for emission quantities. At this point it should be possible to determine extent of controls required and to estimate amount of controlled emissions.

Stage 3 is applicable only when Stage 2 activities do not bring the emissions into compliance. This consists of developing improved control techniques, use of alternative control methods and/or further process modifications. Stage 3 is complete when the amount of controlled emissions, utilizing the improved procedures, is within compliance.

2.1.4 Required Limits

It may not be possible to approach the emission control needs of the emission assessment procedure suggested in the previous section. Under the current requirements of the Clean Air Act, reasonably available control technology (RACT) is specified for existing sources in attainment areas. New sources must have best available control technology (BACT). In particulate nonattainment areas, lowest achievable emission rate (LAER) control is required.

Specifications of control system to achieve RACT, BACT and LAER may vary with person, process and geographic region, and they obviously vary with time. Table 2.1 suggests a possible categorization of some control systems. Note that the control effects are not necessarily cumulative.

Requirements may also exist for operation and maintenance. Operator training, operation procedures and reports may have to be completed, and maintenance procedures, spare parts and schedules may be required. Check with both local and federal requirements for this.

2.2 CONTROL METHODS

2.2.1 Control Options

Fugitive dusts can be controlled by three basic techniques: (1) preventative procedures, (2) add-on equipment and (3) process modifica-

42 FUGITIVE EMISSIONS AND CONTROLS

Table 2.1 Approximate Categorization of Control System Capabilities

Source	Type of Control System					
	RACT		BACT		LAER	
	Control	Efficiency, %	Control	Efficiency, %	Control	Efficiency, %
Unpaved Roads	Wetting agent (water)	50	Wetting agent (other than water)	60–80	Paving and sweeping	85–90
	Speed control	25–35	Drastic speed control	65–80		
			Soil stabilization	50		
			Apply gravel	50		
			Road carpet	80		
Active Storage Piles	Wetting agent (water)	50–75	Wetting agents (other than water)	70–90	Encrusting agents	90–100
	Pile orientation	50–70	Pile orientation	50–70	Tarp cover	100
	Leading slope angle	35	Wind screens	60–80		
Inactive Storage Piles	Vegetation	65	Chemical stabilization plus vegetation	80–90	Tarp cover	100
Transfer Points	Water sprays	35	Wetting agent sprays	55	Enclosure with sprays	90–100
			Fogging sprays	80	Electrostatic-enhanced fogging sprays (EEFS)	80–95
Conveyors	Water sprays	35	Wetting agent sprays	55	Enclosure with sprays	90–100
			Fogging sprays	80	EEFS	80–95
Car Dumpers	Water sprays	35	Wetting agent sprays	40	Enclosure with sprays	85–90
			Fogging sprays	75	EEFS	75–90
Construction Activities	Watering	50	Chemical stabilization	80	Enclosure	90

tion. Preventive procedures include use of chemicals, orientation for low wind erosion, vegetative coverings and wet suppression confinement (natural or otherwise). Wet suppression is used to both knock airborne particles back to the surface and to reduce the formation of the dust. In this regard, for example, most of the emission factor expressions assume dry conditions. Addition of 0.01 in. of water (rain or spray) will usually produce a dust-free condition. Orientation is both with respect to wind direction and pile leading edge slope. Confinement can be partial or complete enclosure of a source by a permanent or temporary structure or by vegetation. This structure is not considered as add-on equipment. Add-on equipment is considered to be a hood and duct collection system with a conventional particulate control device (e.g., cyclones, electrostatic precipitators, baghouses and/or scrubbers). Table 2.2 is a partial listing of some control techniques that could be considered to reduce emissions from various sources as modified and expanded from the matrix of Brookman and Martin [4].

2.2.2 Control Techniques

Fugitive dust controls range from very simple, economical procedures such as watering, to elaborate collection systems with hooding, ducting or control devices. There are economic and practical advantages and disadvantages associated with each technique. For any specific plant, these techniques must be evaluated on specific factors, including need, costs, effectiveness, facility age, design, location, dust characteristics, elevation and temperature of releases, combinations of available control technologies and process modifications. It is also necessary to consider method of waste disposition and availability of satisfactory disposal sites. In disposing of wastes, it may be necessary to control pH and/or otherwise chemically, neutralize the waste; line the waste storage area; eliminate bacteria; prevent spoilage, rodent and vermin infestations; and prevent the occurrence of other types of pollution. The following sections discuss various techniques in general without consideration of the possible interactive effects.

When more than one control technique can be effectively used in series the combined control efficiency can be estimated by a series function. For example, one control that is 50% effective plus a second that is 60% effective yields an overall effectiveness of

$$[1.00 - (1.00 - 0.50)(1.00 - 0.60)]100 = 80\%.$$

44 FUGITIVE EMISSIONS AND CONTROLS

Table 2.2 Control Techniques Guide for Various Sources[a]

Fugitive Emission Source	Chemical Stabilizers	Vegetative Cover	Watering	Windscreens	Wind Barriers/Berms	Plantings	Pile Shaping and Orientation	Paving/Gravel	Sweeping/Cleaning	Reduced Speed	Curbing/Stabilize Shoulders	Operations Change	Reduced Drop Distance	Water Sprays/Foggers	Electrostatic Curtains	Partial or Complete Enclosure	Hooding/Ducting	Covers	Wheel Washes	Foams
Paved Roads	X		X	X	X	X			X	X										
Unpaved Roads	X		X	X	X	X		X		X	X									
Unpaved Parking Lots			X	X	X	X		X		X	X									
Active Storage Piles	X		X	X	X	X	X													
Inactive Storage Piles	X	X	X	X	X	X	X													
Exposed Areas		X	X	X	X															
Construction Sites				X	X	X		X				X								
Conveyor Transfer				X				X				X	X	X	X	X	X	X		
Drop Points												X	X	X	X	X	X	X		
Loading/Unloading					X					X		X	X	X	X	X				X
Vehicle Carryout								X	X										X	
Truck/Rail Spills								X	X											
Crushing/Screening			X					X				X	X	X	X	X	X			X
Waste Sites	X	X	X		X	X		X				X				X		X		
Tilling Operations		X	X									X								
Feed Lots	X	X										X								

[a] Adapted from Brookman and Martin [4].

CONTROL OF EMISSIONS 45

2.2.3 Preventative Procedure Controls

2.2.3.1 Housekeeping

A conscientious program of good housekeeping can almost always result in reduced emissions. This consists of simple, obvious tasks of cleaning up spills, removing accumulations around processing equipment and in general keeping things neat and clean. This usually must be incorporated into the operations and maintenance procedures so that the personnel are aware that this is expected as part of the normal work.

2.2.3.2 Wet Suppression—Water

Use of water to wet down an area or an operation could be an effective, low-cost control technique. This control would be only temporary if the water is not continuously applied. It can result in freezing, runoff, electrical shorting and other problems. Wetting could cause adverse effects: for example, the water could destroy the binder, if present, resulting in *increased* dust.

Once-over, intermittent applications are usually crude soaking spray operations with little attempt to economize on liquid use. About 0.01 in. of liquid applied to the surface is adequate. Spray systems provide refinement, but require the use of nozzles, a liquid pressure adequate to operate the nozzles (a pump may be needed to provide adequate liquid head) and a system to prevent nozzle plugging. This can be achieved by use of clean liquid only, open-type nozzles or use of strainers in or before the nozzles.

Spray Nozzles. Continuous liquid suppression systems usually contain spray systems. The type of system will vary depending on the application requirements. Large quantities of liquid requirements for wetting the surfaces before an operation that creates dust, for example, would require use of nozzles providing large spray droplets. Some of these dust suppression nozzles are listed in Table 2.3. Amount of liquid needed (up to about 0.01 in. on surfaces) and spray area to be covered dictate nozzle type, number and location. Full-cone spray nozzles are usually recommended in these applications. To obtain a nozzle flow rate for an intermediate pressure use the relationship

$$\text{gpm}_2 = \text{gpm}_1 \sqrt{\frac{\text{psig}_2}{\text{psig}_1}}$$

Table 2.3 Typical Nozzles for Dust Suppression and Humidification Requirements

Requirement	Typical Nozzle[a]	Pipe Size in.	Approximate Spray Angle, degrees	Single-Unit Capacity, gpm at pressure, psig			
				3	5	10	20
Dust Suppression	(SS) Fulljet	½	65	1.4	1.80	2.5	3.5
	(SS) Fulljet	1	70	5.4	6.90	9.4	13.0
	(SS) Fulljet	2	95	45.0	57.00	78.0	108.0
	(SS) Fulljet	¼	115		0.74	1.0	1.4
	(SS) Fulljet	⅜	120		1.30	1.7	2.3
	(B) TF	¼	50/60/90/120			1.3	1.9
	(B) TF	⅜	50/60/90/120			3.0	4.2
	(B) TF	½	50/60/90/120			12.0	17.0
	(B) TF	1	50/60/90/120			47.0	67.0
	(B) TF	2	50/60/90/120			178.0	250.0
Humidification	(SS) Fogjet[b]	1	N/A				0.25–48
	(B) L	⅛	90			0.14	0.20
	(B) L	¼	90			1.27	1.72
	(B) P	¼				0.033–1.27	0.046–1.72

[a] (SS) = Spray Systems Co., N. Ave. at Schmale Road, Wheaton, IL 60187; (B) = Bete Fog Nozzle, 34 Wells St., Greenfield, MA 01302.
[b] Multiple.

Theoretical spray coverage at various distances and angles is shown in Table 2.4.

Humidification is applicable for keeping the air moist and thereby keeping the dust humidity high. This causes coalescence and coagulation and maintains a high cohesive force. Typical nozzles for fogging are shown in Table 2.3. Amount of liquid will depend on amount and type of dust and moisture content of the dust. Number and location of nozzles will vary with amount of liquid required and the space to be sprayed. Note that some fogging nozzles are supplied as clusters with several in a group, and therefore spray angles are not shown. All humidification-type nozzles require higher pressures and liquid strainers.

Lower liquid requirements are being achieved by the use of electrostatically charged fog atomization systems. One system currently being tested by the EPA is a spinning cup fog thrower, which can be used on moving vehicles such as road sweepers and front end loaders [5]. It appears to be well suited for fine particle control also. Other studies [6] show that water sprays in certain rock processing applications were 30 to 40% effective in reducing dust. Through use of a fog charged by a 7 kv dc induction ring, dust control for the same application increased to 40 to 70%. No difference in effects with polarity of charge were observed. These systems are commercially available. One supplier is The Ritten Corp., Ltd (40 Rittenhouse Place, Ardmore, PA 19003).

2.2.3.3 Wet Suppression—Chemicals

Certain benefits and disadvantages can be obtained by the additions of chemicals to the water used for suppression. Chemicals available include bituminus-asphaltic compounds, polymers, resins, surfactants, latex, emulsion, calcium lignosulfonate and others. Data obtained from PEDCo [7] and TRC [8] are presented as Table 2.5. Included are typical

Table 2.4 Theoretical Nozzle Spray Coverage,

Spray Angle, degrees	Coverage in in., at distance from nozzle, in.					
	4	8	12	18	24	30
50	3.7	7.5	11.2	16.8	22.4	28.0
60	4.6	9.2	13.8	20.6	27.7	34.6
65	5.1	10.2	15.3	22.9	30.5	38.2
70	5.6	11.2	16.8	25.2	33.6	42.0
90	8.0	16.0	24.0	36.0	48.0	60.0
95	8.7	17.5	26.2	39.3	52.4	65.5
120	13.9	27.7	41.6	62.4	83.2	104.0

Table 2.5 Chemical Suppressants for Fugitive Dust Control

Company/Address	Product Name/ Product Type	Cost[a]	Uses/Comments	Density, Dilution and Application Rates
Dow Chemical Co. 2020 Dow Center Midland, Mich.	XFS—4163L Styrene-butadiene	55-gal drums 1 drum $2.65/gal 25 drums $2.15/gal Bulk $1.90/gal	Mulches such as straw, wood cellulose fiber and fiberglass. Used to prevent wind loss of mulches during stabilization periods such as reseeding periods.	8.5 lb/gal 40 gal XFS—4163L: 360 gal water 400 gal/acre
Witco Chemical Corp. Golden Bear Division Post Office Box 378 Bakersfield, Calif. 93302	Coherex Cold water emulsion of petroleum resins	55-gal drums 1–10 drums $0.65/gal >10 drums $0.63/gal Bulk $0.38/gal	Unpaved haul roads and stockpiles. Can be used around human or animal habitats—very clean—no heat required. Can be stored for 12 mo or longer. Must be protected from freezing—unless freeze stable type is used. Can be spread through any type of equipment used to spread water.	8.33 lb/gal 1:4 dilution, 1–1.5 gal/yd^2 for parking lots and dirt roads. 1:7 dilution 0.5 to 1 gal/yd^2 for thin layer or loose dirt, light traffic, service roads. 1:10 dilution for aid in packing surface
	Semipave Cold asphalt cutback with antistrip agent	55-gal drums 1–10 drums $0.68/gal >10 drums $0.64/gal Bulk $0.39/gal	Penetration of unpaved areas—low traffic volume roads—parking lots etc. Can be handled without heat if ambient temperature is 50°F or higher.	250 gal/ton 0.6 to 0.8 gal/yd^2
American Cyanamid Wayne, New Jersey 07470	Aerospray 52 binder	55-gal drums 1–4 drums $0.69/lb 5–11 drums $0.66/lb 12–22 drums $0.63/lb 23–53 drums $0.61/lb >53 drums $0.59/lb Bulk $0.55/lb	Seed membrane protection, excavation, construction, slope stabilization	8.8 lb/gal 2:1 1 gal/100 ft^2

Company	Product	Pricing	Applications	Properties
E. F. Houghton & Co. Valley Forge Tech. Center Madison & Van Buren Ave. Norristown, PA 19401	Surfax 5107	55-gal drums 1–4 drums $4.44/gal 5–9 drums $4.41/gal 10–39 drums $4.38/gal >39 drums $4.35/gal	Coal loading, quarries, cement plants, crushers, sintering plants	8.5 lb/gal 1:1000 or higher
	Rezosol 5411-B Polymer	55-gal drums 1–4 drums $0.415/lb 5–9 drums $0.41/lb 10–39 drums $0.405/lb >39 drums $0.40/lb	Storage piles, railcars, road sides	8.75 lb/gal 1:30 40 gal/1000 ft^2, recommended 2 applications
Monsanto 800 N. Lindbergh Blvd. St. Louis, MO 63166	Gelvatol 20-90 Polyvinyl alcohol resin	50 lb/bags 500 lb $0.905/lb 2,000 lb $0.80/lb 10,000 lb $0.77/lb 30,000 lb $0.74/lb >30,000 lb $0.725/lb	Surfactant and protective colloid in emulsion polymerization	30–40 lb/ft^3 10 to 20 wt %
	Gelva Emulsion S-55 Polyvinyl acetate homopolymer	55-gal drums 1–3 drums $0.27/lb 4–19 drums $0.26/lb >19 drums $0.25/lb Bulk $0.205/lb	Adhesives	500 lb/55 gallon drum 1 wt %
Air Products & Chemicals, Inc. 5 Executive Rd. Suedesford Road Wayne, PA 19087	Vinol 540 Polymer (water-soluble)	50-lb bags 500 lb $0.80/lb 2,000 lb $0.77/lb 10,000 lb $0.74/lb 32,000 lb $0.725/lb 120,000 lb $0.72/lb	Two grades: (1) soluble in water (washed away with rain), (2) relatively insoluble in water.	1 to 7 wt % Slurried in cold water or heated to ensure complete mixture in solution
Union Carbide Corp. West St. & Madisonville Rd. Cincinnati, Ohio 45227	DCA-70		Stabilize steep grades, tailings ponds. Not for vegetation growth.	9.25 lb/gal 2:1
Enzymatic Soil of Tucson 6622 N. Los Arboles Cr. Tucson, Arizona 85704	Enzymatic SS	55-gal drums $7.60/gal	Hold down dust on haul roads, tailings, stockpile. Will retard growth of weeds or plants. Seal lakes, stock tanks, stabilize odors around stock pens.	8.34 lb/gal 1:1000 1000 gal/20 to 30 yd^3

FUGITIVE EMISSIONS AND CONTROLS

Table 2.5, continued

Company/Address	Product Name/Product Type	Cost[a]	Uses/Comments	Density, Dilution and Application Rates
Asphalt Rubberizing Corp. 1111 S. Colorado Blvd. Denver, Colorado 80222	Peneprime Low-viscosity, special hard-base asphalt cutback	10,000-gal lots $0.45/gal	Control of wind, rain, or water erosion of soils. Applied to roads and streets to allay dust and stabilize surface to carry traffic. Does not allow seed germination. Very light applications (0.2–0.4 gal/yd^2) may accelerate seed germination due to warming of black surface. Applications above 0.4 gal/yd^2 inhibit plant growths through hardness and toughness of the crust formed. Plant growths through the crust may be further inhibited by addition of several oil-soluble sterilants. Sterilants kill plant as it emerges. The material may be applied at temperatures as low as 75°F by conventional asphalt distribution equipment.	0.85 S.G. dust abatement, 0.2 gal/yd^2 erosion control, 0.5–1.0 gal/yd^2
Johnson-March Co. 3018 Market St. Philadelphia, PA 19104	Compound-MR (regular)	55-gal drums 1–3 drums $6.00/gal 4–11 drums $5.00/gal >11 drums $3.35/gal	Usually used with a spray system or storage piles, conveying systems.	1:1000 water applied as needed
	Compound-SP-301	1–4 drums $1.80/gal 5–9 drums $1.75/gal 10–44 drums $1.70/gal >45 drums $1.65/gal	Used on haul roads, parking lots, stabilizing cleared areas, aid in vegetation growth.	1 gal/100 ft^2 ± depending on conditions. Application lasts 6 mo to 1 yr
	Compound-MR (superconcentrate)	$6.75/gal	Same as Compound-MR (regular)	1:3500 water

CONTROL OF EMISSIONS 51

Company	Product	Price	Description	Application
Grass Growers P. O. Box 584 Plainfield, NJ 07061	Compound-SP-400	1–4 drums $3.50/gal 5–9 drums $3.40/gal 10–44 drums $3.30/gal >44 drums $3.20/gal	Same as Compound SP-301	Same as Compound SP-301 Application lasts 1 to 5 yr
	Coal tarp	$0.75–$1.00/gal	Designed for use in coal industry: coating over rail cars, trucks to prevent transportation losses etc. Prevents seed germination.	
	Tarratack-1	$2.25/lb	Mulch binder used for stabilizing any type of grass to be grown.	5 lb: 250 gal water, mixed with wood fiber mulch (40 lb/acre) 5 lb: 150 gal water, mixed with hay or straw (40 lb/acre) Mixed with hay or straw 40 lb/acre
Dubois Chemical Dubois Tower Cincinnati, Ohio	Tarratack-2	$2.75/lb	Same as Tarratack-1	Mixed with wood fiber only 1–2 lb/1000 gal
	Tarratack-3 Floculite 600	$3.25/lb 100 lb $2.81/lb 1000 lb $2.74/lb	Same as Tarratack-1 Used in wastewater treatment from mines. Also helps keep down dust on haul roads.	
Mona Industries, Inc. 65 E. 23rd St. Paterson, NJ 07624	Monawet Mo-70E	500-lb drums 1–50 drums $0.455/lb Bulk $0.385/lb	Used in coal industry as dust suppressant.	0.1% in water, must be reapplied when water evaporates 8.2 ± 0.1 lb/gal
AMSCO Division Union Oil Co. of California 14445 Alondra Blvd. La Miroda, CA 90638	Res AB-1881 Styrene Butadiene		Soil stabilizer, particularly in conjunction with wood fiber mulches. Free pumping in conventional hydroseeding equipment. Not to be applied in soils with pH less than 6.0.	
Witco	Coherex		Oil-latex emulsion Control can be 50–80%	1 part emulsion 5 parts water Bladed into gravel
ARCO Mine Sciences 1500 Market Street Philadelphia, PA 19101	AMS 2400 and AMS 2400 HP		Elastomeric asphalt emulsion chemical dust suppressant on unpaved road	Bladed into gravel

Table 2.5, continued

Company/Address	Product Name/Product Type	Cost[a]	Uses/Comments	Density, Dilution and Application Rates
Deter Co., Inc. 8 Great Meadow Lane E. Hanover, NJ 07936	Deter Pressure Foam		Low- or high-expansion foam reduces water spray rates to less than 0.5 gal per ton treated material and increases control by 20% over water only.	
ARCO Mine Sciences as above	AMS 4200 Foam and AMS 4060		Suppressant for equipment dust from conveyors, crushers, etc.	
Johnson March Corp. 3018 Market Street Philadelphia, PA 19104	Chem Jet MR		Proprietary chemical	1 part to 1000 parts water Use up to 2.5 gal mix per ton coal.

[a] As of 1976.

applications, application rates and costs (as of 1976). Note that addition of these chemicals to water in a spray system can change both density and viscosity of the fluid. Correct the spray values in Table 2.3 for density by

$$\text{gpm}_2 = \frac{\text{gpm}_1}{\sqrt{\rho_2}}$$

where ρ = fluid specific gravity in g/cm^3.

Effects of viscosity changes are most easily determined empirically.

The spray chemicals can react with the material handled, causing unexpected results. This should be determined in advance by compatibility testing.

2.2.3.4 "Dry" Suppression

Dust in the atmosphere surrounding an operation can be suppressed by use of sonic and electrostatic devices. The systems utilize water, but the amount is so low it evaporates and does not wet the product. It does increase humidity, however. The dry sonic system of Sonic Development Corporation (3 Industrial Ave., Upper Saddle, NJ 07458) generates very finely atomized water droplets which cause the dust to grow in size, agglomerate and fall. One 500-ton/hr (tph) asphalt mix plant uses sonic agglomerator nozzles to suppress dust from crushing, screening and transfer operations. The plant uses only 400 gal/day of water and the nozzles cost 20% of the cost of a baghouse. This technology has potential; however, it has not been demonstrated as adequately as other systems have.

2.2.3.5 Soil/Waste Stabilization

Dust can be suppressed by compression of the materials. In addition, chemical binders can be added. This is often very expensive—but it could be the cheapest alternative. Chemical companies, such as NALCO Chemical Company (2901 Butterfield Road, Oak Brook, IL 60521), can recommend and supply specific binders for individual uses. In addition, others, such as Dravoe (Dravoe Corporation, One Oliver Plaza, Pittsburgh, PA, 15219), use proprietary chemicals, but will either supply the chemicals or provide a service to treat the material. A partial listing of soil stabilizers is given in Table 2.6. Addition of more chemical than

Table 2.6 Soil Stabilizing Chemicals and Control Efficiencies

Dust Suppression Chemical (in water as listed)	Typical Control Efficiency, %
T-Det 1:4	76.0
CaO 1%	2.8
$CaCl_2$ 2%	33.8
Cements 5%	26.8
Coherex 1:4	97.2
Dowell Chemical Binder 1%	70.4
Dowell Chemical Binder 2%	97.2
1% $CaCl_2$, in 1:5000 Dustrol "A"	15.5
1% CaO in 1:8 Coherex	31.0
1% CaO, 1:3000 T-Det in 2% Dowell Chemical Binder	98.6
1% CaO, 1% $CaCl_2$, 1:4000 Dustrol "A" + 2% Dowell Chemical Binder	98.6
Elastomeric Asphalt Emulsion	86.0–90.0
Latex-Oil	50.0–80.0

shown does not necessarily mean improved dust suppression. In some cases, more chemical causes reduced control.

2.2.3.6 Confinement

Partially or completely closing off the source from the atmosphere to prevent the wind from entraining the dust can effectively reduce emissions where this is possible. Examples where this may be practical include transfer points, conveyor belts, storage areas, relatively inactive piles, and truck/railcar loading and unloading. Frequently, confinement plus suppression is an optimum combination for effective, economical dust control. Confinement techniques are closely related to wind control techniques which are discussed in the following section.

2.2.3.7 Wind Control

Preventing the wind from entraining the dust particles can be accomplished by keeping the wind from blowing over the material. This can be done by confinement or by wind control. Tarps to cover piles are a wind control/confinement procedure for storage piles, trucks, railcars and other sources.

Truck Tarps. Simple covers can be tied onto the truck or fitted systems such as the Aero Industries "Crank N Go" tarp can be used (3010 W. Morris St. Indianapolis, IN 46221). These systems include a cranking

CONTROL OF EMISSIONS 55

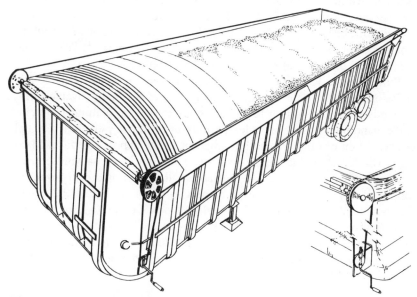

Figure 2.1 Complete truck tarp, tarp movement and hold-down system.

mechanism, guides, crossbows and tarp, as shown in Figure 2.1. Proper use can provide nearly 100% protection from dust.

Civil Engineering Fabrics. Fabrics engineered for both top and bottom cover can be more than 50% effective in control of both fugitive dust and waterborne fugitive emissions [2]. They are useful for storage piles, ground stabilization (unpaved roads), erosion control and in construction and maintenance. They are used on top, for example, with piles, and are over 90% effective. They can be used as "carpets" on unpaved roads where they are placed on the existing soil/gravel bed and then covered with coarse aggregate to effect dust reductions of 50% or more. Underneath the material, they can completely prevent drainage. Some suppliers of linings and other fabrics are Johns Manville Corporation, Box 5705, Denver, CO 80217; Randustrial Corporation, 13311 PL Union Ave., Cleveland, OH 44120; Diamonite Products Manufacturing Co., W. McConkley St., Shreve, OH 44676; and Federal Bentonite Division, 1002 Greenfield Road, Montgomery, IL 60538. Other addresses can be found in the Chemical Engineering Equipment Buyer's Guide [9].

Encrusting. Storage piles and moving material surfaces can be protected by a surface encrusting. This may be by physical action of the material itself (e.g., flame heating surface of sulfur piles to produce a molten surface that solidifies). Encrusting agents can be added to the

56 FUGITIVE EMISSIONS AND CONTROLS

surface. ARCO Mine Services (1500 Market St., Philadelphia, PA 19101) produces an AMS 2000 binder that forms a protective coating that is not affected by sunlight and is highly resistant to wind and rain. The same company provides AMS 2010, which is somewhat flexible and can be used on material being transferred or on piles.

Windscreens. Reduction in wind speed, and therefore reduction in emissions, can be achieved using one or more of the many diverse forms of windscreens in addition to enclosing the dust area. Natural forms include windbreaks, such as piles or hills upwind, as well as trees, tall grasses, grain or other vegetation adjacent to the exposed surface. Man-made windbreakers include devices made and marketed as windscreens.

Commercial windscreens are portable and can be placed in front, on top, or at any desired position in respect to a source. Figure 2.2 shows a commercial screen supplied by Julius Koch Co. (P.O. Box A-995, New Bedford, MA, 02741). This is constructed of 8-ft modular units interconnected to form a continuous-length fence that may be relocated as needed. The specific application shown is on top of a coal storage pile at a 1600-MW electric utility.

Figure 2.2 Commercial windscreen. (Courtesy Julius Koch, Inc., New Bedford, MA.)

A windscreen on top of a pile, as shown in the figure, is equal in effect to constructing a windscreen the height of the pile plus the height of the modular screen unit. This precludes the need to construct enormous screens to protect a pile.

Windscreens have been tested and found to be effective for aggregate pile dust reduction. Wind speed reductions as a function of fence height are shown in Figure 2.3a as a side view profile and in Figure 2.3b as a plan view profile [10]. Note that the 50% wind isotach (line denoting constant wind speed) extends about 12 wind screen heights downwind. These values were obtained using a 16-ft (4.9-m) screen with 50% porosity and a 38-mph wind speed on a mown field. Doubling the height would double the downwind distance. Figure 2.3 shows the edge effects and also the turbulent eddies developed at the edges where wind speeds are greater than incoming wind speed. Measurements at lower incident wind speeds are given in Figure 2.4. This shows that as wind speed increases the isotachs increase more quickly with distance. The effect of screen porosity is shown in Figure 2.5. The net overall effectiveness of these types of screens in reducing fugitive emissions is 80% [10] although some earlier work [7] reports screen control efficiency as "very low." Carnes and Drehmel [10] estimate that a 60×10^6-ton coal pile 75 m long by 30 m wide that would normally lose 15.1 tons dust per yr would only loose 3.1 tons/yr using a 30-m-long screen which costs \$7200 and has a 10- to 11-year life. This does not include installation charges.

Pile Orientation. Emissions can be influenced for a given material by pile geometry and geographical and meteorological conditions, as discussed in Section 1.6.2.5. It has been shown that emissions can be reduced by decreasing the influence of the wind. Emissions can be further minimized by:

1. reduction of K_4 by reducing the leading edge slope of the pile,
2. reduction of K_5 by aligning the longitudinal axis of the pile with the predominant wind direction, and
3. reduction of the local wind velocity over the pile.

Field and wind tunnel testing indicates that for a pile leading edge slope of 30° (as shown in Figure 1.9), the K_4 factor is approximately 1.5. Reducing this slope to 10° will reduce the K_4 factor to approximately 1.0. It should also be noted that a reduction of this angle will require more storage area for the pile and may not be practical for all situations.

The value of K_5 may also be minimized by aligning the pile longitudinal axis with the predominant wind direction, as shown in Figure 1.10.

58 FUGITIVE EMISSIONS AND CONTROLS

Field test data show that storage pile emissions are proportional to the projected pile frontal area. Worst case emissions occur when the pile is aligned 90° to the predominant wind direction.

A practical consideration for the above criteria is to determine the direction of predominant wind and align the pile in an optimum direction accordingly. This may be approached by defining the wind velocity mag-

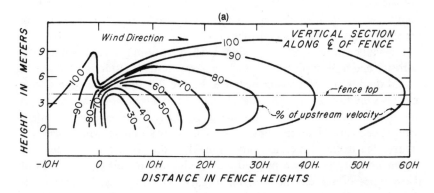

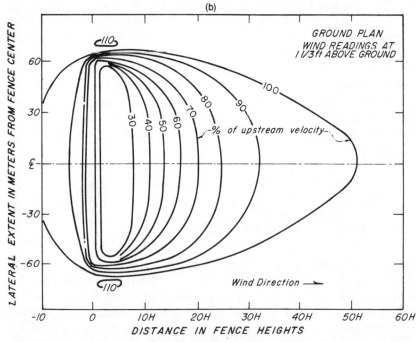

Figure 2.3 Wind velocity patterns above a mown field with a windscreen [10]. (a) Side view profile. (b) Plan view profile.

CONTROL OF EMISSIONS 59

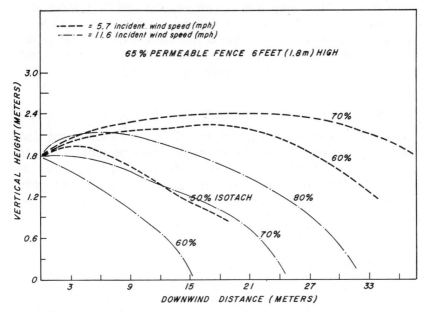

Figure 2.4 Measured velocity reductions from windscreen [10].

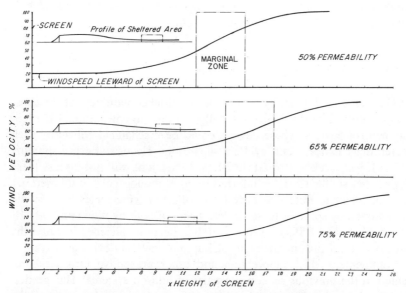

Figure 2.5 Leeward flow field for variations in windscreen permeability.

nitude/direction frequency and determining the minimum "integrated emissions" direction. This should provide minimum emissions on a long-term basis. However, the reduction of free stream wind velocity in the vicinity of the storage pile, as discussed above, is the most important and efficient method of reducing storage pile emissions.

2.2.3.8 Operation and Maintenance (O & M)

This has been mentioned in Section 2.2.2.1 on Housekeeping, but deserves special attention because of the numerous beneficial results of good O&M. Good O&M includes prevention as well as elimination of many types of fugitive emissions, such as process upsets, improperly operated equipment, defective equipment, poor understanding of the process and control procedures and poor housekeeping.

2.2.4 Process Modification Controls

This section is included simply as a reminder that it may be more practicable to change a process and/or operation to reduce emissions than to try to control the emissions. For example, a pneumatic conveyance system could eliminate the emission problems of a conveyor belt, which might require any or all of the preventive procedures (enclosures, windscreens, wet suppression, housekeeping of spilled materials, stabilization of material) and add-on equipment.

2.2.5 Add-On Equipment Controls

This consists of a facility to provide an initial containment of the emissions, followed by ducting of the collected dust to conventional particulate control devices. These devices could be mechanical collectors such as settling chambers and cyclone separators, electrostatic precipitators (ESPs), wet scrubbers or fabric filters. (This book will not discuss these devices per se; the reader is referred to other monographs in this series.) It is sufficient to note that very high collection efficiencies can usually be obtained by some of these devices (e.g., 99.9% by filters and 99.6% by scrubbers and ESPs) but the overall collection efficiency includes the effectiveness of the confinement, pickup and control device as a system.

Confinement, dust suppression and other preventive measures presented in the previous section are parts of add-on systems. This section deals mainly with the pickup and transfer requirements for bringing the dust from the confinement area to the collector.

2.2.5.1 Dust Capture

Many of the procedures recommended by the Committee on Industrial Ventilation [11], other industrial hygiene publications, and publications such as the PEDCo EPA guide [7] are applicable to fugitive dust pickup procedures. Four basic requirements exist:

1. A minimum face velocity is required, depending on the dust mean particle size and density, to minimize escape losses.
2. An adequate air volumetric flow rate is needed to carry the dust away.
3. A minimum duct velocity is necessary to prevent the dust from settling in the duct.
4. Pickup hood and duct geometry and arrangement must meet the system needs.

The third requirement is simplest. Gas velocities should be about 50 to 60 ft/sec to prevent dust settling and to minimize dust abrasion in the system.

Face Velocity or Capture Velocity. This is the velocity of the gas moving into the hood necessary for the hood to capture the dust. Some typical face velocities for dust pickup are given in Table 2.7. For dusts with densities significantly greater than 2 g/cm^3, it would be advisable to multiply the values in Table 2.7 by ½ the density.

In addition to the face velocities in Table 2.7, Dalla Valle [13] suggests a procedure for estimating centerline velocity, v in ft/min, at a distance x in feet from a hood by:

$$v = \frac{Q}{10x^2 + A}$$

where Q = air volumetric rate, cfm
A = hood opening area, ft^2

This is applicable for round and rectangular hoods of essentially square configuration where x is within 1.5 times the equivalent diameter. (Equivalent diameter equals 4 times the cross section area divided by the "wetted" perimeter.)

Hood Design. Air flow rate and velocity into a hood must be adequate for the task intended. It is important to account for extraneous influences such as:

62 FUGITIVE EMISSIONS AND CONTROLS

- wind currents (room currents can be 50 ft/min or more),
- motion of device, moving vehicles or people,
- thermal currents,
- eddy effects,
- area and shape of dust confinement required,
- particle physical properties, and
- dust concentration.

Some of these factors have been discussed, but many can best be established by empirical measurements. Figure 2.6 shows some hood arrangements and air flow requirements based on air inlet centerline velocity to the hood. Note that if a flange is added at the inlet portion of the hood, the volumetric flow rate is reduced to 75% for the regular (rectangular or round) and slot-type dust pickup hoods. If the material is hot, the gas flow rate should be increased by $\Delta t^{0.42}$, where Δt is the temperature difference between the hot source and the surrounding atmosphere in °F, according to Henson [15].

The canopy hood velocities typically range from 100 to 500 (and possibly to 2000) as given in Table 2.7. The slope of the hood should be >45° as noted and the dimensions of the confinement below the hood should be no greater than 0.4D where D is the height above the confinement. Enclosing the hoods on as many sides as possible makes them more effective and reduces the gas flow requirements accordingly.

Transport Velocity. This must be adequate to prevent dust dropout but low enough to prevent excessive abrasion. Velocities as low as 40–50 ft/sec may be adequate for light dusts, but for dusts heavier than about 1 g/cm^3 higher velocities are required. Table 2.8 summarizes some typical values. A is area in square feet.

Transport Ducts. Duct sizes must be tapered to increase or decrease transport cross-sectional area as inlets are added or exhausts are re-

Table 2.7 Typical Dust Pickup Face Velocities for Low-Toxicity Materials [11, 12]

Operation	Face Velocity, ft/min
Hopper Ventilation	150
Mixer	200–300
Bag Filling Station	450–500
Belt Conveyor	150–200
Conveyor Transfer	100–200
Conveyor Loading	200–500
Crushing, Grinding	500–2000

CONTROL OF EMISSIONS 63

TYPE	HOOD ASPECT RATIO, $\frac{W}{L}$	TYPICAL AIR VOLUME
REGULAR 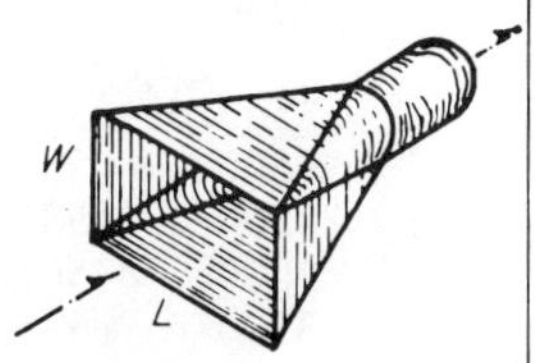	> 0.2	$Q = U(10X^2 + A)$
SLOT 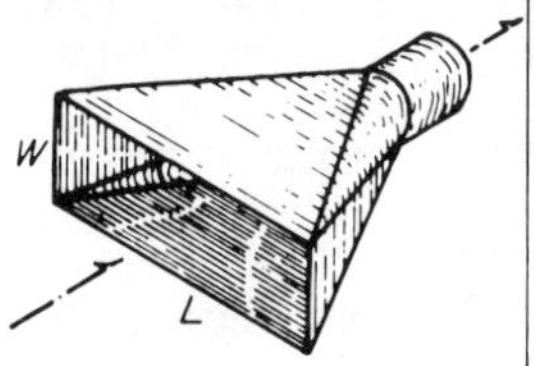	≤ 0.2	$Q = 3.7 L U X$
BOOTH 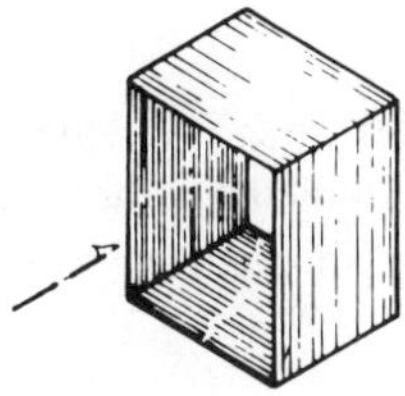	VARIOUS	$Q = UA = UWH$
CANOPY (45° min.) 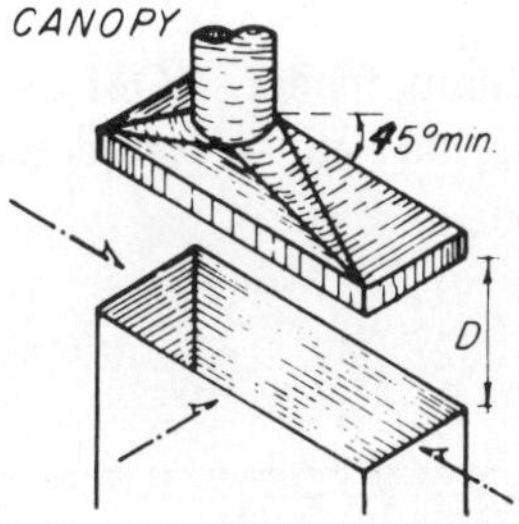	VARIOUS	$Q = 1.4 PDU$ P = Perimeter of work D = Height above work

Figure 2.6 Hood arrangements and air flows [13,14].

moved, respectively, as shown in Figure 2.7. The duct tape should be 1 in. in diameter per every 5 in. in length. Inlets should be staggered as shown in Figure 2.7 and should not enter directly opposite each other

64 FUGITIVE EMISSIONS AND CONTROLS

Table 2.8 Typical Transport Velocities and Air Flow Rates

Operation	Air Flow Rate, cfm[a]	Transport Velocity, ft/min	(ft/sec)
Bin, open	200 A	3500	(58)
Conveyor, belt	350 to 500 times w (w = belt width, >1 ft)	3500	(58)
Crusher	200 A	3500	(58)
Elevator, bucket	100 A	3500	(58)
Mixers	200 times opening areas	3000	(50)
Rock, drilling	60–200	3500	(58)
Screens	50 times screen area or 200 A, whichever is greater	3500	(58)
Weighing, packing	Up to 150 times dust processing area	3000	(50)

[a] A is area in ft².

if possible. Inlet duct angles of no more than 30° from the main duct are recommended.

2.2.6 Example Arrangements

Two coal handling facilities are used as examples to show system variations and requirements for each. The second system uses chemical wet suppression plus a baghouse to control emissions, in contrast to the first system which is a dry only system.

2.2.6.1 *Coal Handling Facility—Dry Collection*

Figure 2.8a shows the basic system of truck dump, feeder, ROM conveyor, crusher and loadout conveyor. The car loadout is completely

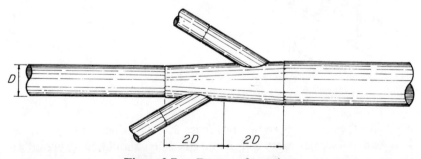

Figure 2.7 Duct configuration.

enclosed and vented through the loadout conveyor. All systems are ducted to the baghouse, and size assumptions are implicit in the calculations.

Air Volume Requirements

Hood	Air Flow, cfm	
A	(150) (20 ft × 20 ft) =	60,000
B	(500) (54/12) =	2,250
C	(200) (15 ft × 15 ft) =	45,000
D	(500) (36/12) =	1,500
Baghouse Total		= 108,750

The cost of a baghouse system plus a 300-hp fan/motor would be about $450,000 (in 1980 dollars). Operating costs would be mainly due to gas pressure drops and associated baghouse filter costs.

Duct Requirements at 3,500 ft/min

No.	Area, ft^2	Diameter, ft
1	17.14	4.7
2	0.64	0.9
3	17.79	4.8
4	12.86	4.1
5	30.64	6.3
6	0.43	0.7
7	31.07	6.3

2.2.6.2 Coal Handling Facility—Wet Chemical Suppression

This system is the same as the previous system except that wet chemical suppression is used to reduce emissions and only the discharge from the crusher and load-out conveyor are vented to the baghouse, as shown in Figure 2.8b. The baghouse is now only about 1.5% of the original size; however, there are the added costs of chemicals, spray system and water. Chemicals would cost about 0.5¢ per ton of coal handled [16].

Air volume requirements for the single hood are (500) (36/12) = 1500 cfm.

Course sprays are used at the dump and feed points and fogging sprays are used at the crusher. This system would cost about $100,000 (in 1980 dollars).

Additional factors to be considered in this type of facility would vary with the operations involved. For example, control of the coal unloading

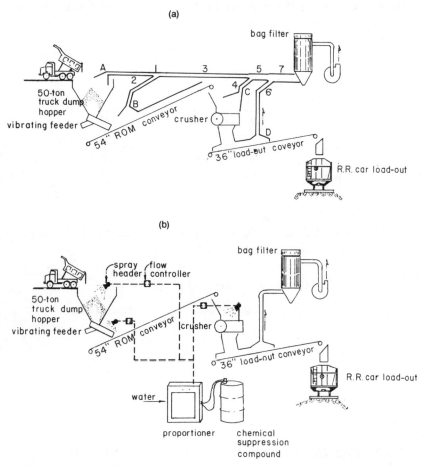

Figure 2.8 Coal handling facility. (a) Dry collection. (b) Wet collection.

could consist of one or more of the following procedures. Railcar and truck unloading—control by:

- drive-through enclosure with doors at both ends
- exhaust of enclosure to dust removal equipment
- exhaust air from below grating of receiving hopper to removal equipment
- choke feed receiving pit (hopper car and hopper truck)
- unload with screw conveyor (boxcar)
- wet suppression (water, chemicals)
- pneumatic unloading system

Barge and ship unloading—control by:

- enclosed top of clamshell bucket with transparent material for maintenance of closure seals and teeth on bottom of bucket
- enclosed shoreside receiving hopper
- exhaust enclosed shoreside receiving hopper to dust removal equipment

2.3 CONTROL BY INDUSTRY

This section discusses fugitive emissions and control techniques and their effectiveness in typical operations and processes. The information provided is only suggestive, and an actual situation must be evaluated and modified on a case-by-case, operation-by-operation basis. For those industries not listed, use one of the related industry examples as a guide. Supplemental emission factors and other data are included in this section; however, refer to Chapter 1 for the basic emission factor assumptions and conditions.

2.3.1 Agricultural Operations

Essentially all agricultural emissions are fugitive, as shown by the listing in Table 2.9 [17]. They include both particulate and gaseous fugitive emissions.

2.3.1.1 Emissions and Controls

Field Operations. This includes the small quantities of emissions from tractors, trucks and airplanes used in the field work. Estimate these emissions as 0.03 g particulates, 1.2 g hydrocarbons (HC), 20 g carbon monoxide (CO), and 6 g nitrogen oxides (NO_x) per gallon of fuel. Emissions are controlled by use of factory-provided mechanisms and by good vehicle operation and maintenance procedures. In contrast, large quantities of particulates are released during tilling operations, as discussed in Section 1.5.2. Control procedures include "no till" operation, low vehicle speeds, working the soil when moist, working the soil during times of low wind speed and attempting to provide vegetative cover as soon as practicable. Field emission quantities are shown in Table 2.10. Vegetative cover, e.g. by hydroseeding, could cost from $300 to $500/ acre (in 1980 dollars).

68 FUGITIVE EMISSIONS AND CONTROLS

Table 2.9 Significant Direct Agricultural Related Air Pollution Sources

Category	Particulate	NO_x	HC	CO	Odor	Visibility
Plant-Related						
Field Operations:						
cultivation	F	F	F	F		F
planting	F	F	F	F		F
harvesting	F	F	F	F		F
fertilization	F	F	F	F	F	F
pest/insect control	F	F	F	F	F	F
Growth:						
pollen	F		F		F	F
spores	F		F			F
mechanical disintegration	F		F			F
Decal:						
aerobic	F		F	F		F
anaerobic			F	F	F	F
Product Storage	F		F		F	F
Animal-Related:						
Feed Lot/Pasture	F		F		F	
Manure	F		F		F	
Slaughter/Processing			S		S	
Transportation-Related:						
Farm Roads	F	F	F	F		F
Repairs	F					
Other:						
Buildings						
office shops	S	S	S			
barn			S		S	
Heating/Steam Generation	S	S	S			S
Fires:						
field/forest	F	F	F	F	F	F
smokehouses/roasting, etc.	S	S	S	S	S	S
incinerators	S	S				
orchard/grove heaters	F	F	F			

[a] F = fugitive, S = stack.

Spraying. The spraying of liquids such as fertilizers, insecticides, pesticides and other chemicals creates liquid particulate pollutants, e.g., hazardous chemicals, HC or other materials. This type of fugitive particulate emission would not normally travel very far. However, fine mists and evaporated material could travel significant distances. To reduce emissions, low-pressure spray nozzles (e.g., ≤ 20 psig) can be used to

Table 2.10 U.S. Agricultural Emissions in 10^3/ton for 1976 [18]

State	Tilling	Wind Erosion	State	Tilling	Wind Erosion
AL	35.0	70.0	NE	3907.0	7280.0
AK	0.3	22.0	NV	400.0	253.0
AZ	535.0	1189.0	NH	20.0	4.5
AR	338.0	158.0	NJ	7.3	25.0
CA	2422.0	6669.0	NM	52.0	5412.0
CO	2104.0	4663.0	NY	91.0	212.0
CT	2.1	8.7	NC	37.0	285.0
DE	4.2	25.0	ND	7256.0	80667.0
FL	12.0	330.0	OH	322.0	7449.0
GA	28.0	239.0	OK	856.0	3396.0
HI	9.4	197.0	OR	123.0	329.0
ID	1076.0	2601.0	PA	111.0	260.0
IL	1150.0	2763.0	RI	0.3	0.3
IN	321.0	873.0	SC	16.0	133.0
IA	1779.0	8510.0	SD	3968.0	30926.0
KS	4034.0	21464.0	TN	159.0	47.0
KY	84.0	238.0	TX	1172.0	5384.0
LA	116.0	85.0	UT	295.0	424.0
ME	4.2	20.0	VT	9.4	27.0
MD	15.0	46.0	VA	45.0	111.0
MA	2.4	27.0	WA	51.0	1226.0
MI	132.0	259.0	WV	22.0	29.0
MN	1721.0	15304.0	WI	511.0	1574.0
MS	85.0	59.0	WY	452.0	1161.0
MO	997.0	2176.0	Totals	39,012.0	229,655.0
MT	2103.0	15021.0			

minimize the generation of fine particles. Also, spray during periods of low wind velocities.

Growing Plants. As plants grow they give off odorous hydrocarbon emissions which often have very pleasant odors. Nevertheless, they are hydrocarbons and they do create problems. No control for this type of emission exists. Particulate emissions from vegetation include pollens, spores and broken vegetation, all of which can cause common respiratory problems. These materials are mostly 10 to 100 μ in size and can be controlled by early cutting practices.

Decay. Aerobic decay (in the presence of air) of plants produces hydrocarbon emissions, but these are minor. *Anaerobic* decay (in the absence of air) can result in the production of large amounts of flammable methane (CH_4) and carbon monoxide (CO). These gases can be captured using stand pipes connected to pressure "storage vessels" (this includes earth-covered landfill). These gases are useful as fuels.

Grain Dusts. (See also Section 2.3.5) Harvesting and grain handling can produce large quantities of particulate and HC emissions. Harvesting could release over 10 lb of dust per ton of product. Transfer, conveyance and cleaning could each produce respectively about 2, 3 and 7 lb of dust per ton of product handled. These emissions can be cut by 70%, while recovering valuable product, by use of closed containers equipped with simple cyclone separators. This would require a low-capacity blower to introduce the entrapped air and dust into the cyclone at an inlet velocity of 50 to 90 ft/sec, and at a flow rate of 300 D^2 (ft^3/min), where D is the cyclone body diameter in feet.

Odors. Animal-related air pollution emissions are probably the most infamous for odorous emissions. These are mainly gaseous hydrocarbons but may include windblown dust as well. Characteristic barnyard odors are a nuisance at very dilute levels (e.g., at parts per billion by volume of gases). Many of these emissions can be controlled by good farm management practices. In addition, adsorbents can be used to react with/adsorb the odors. These include the use of chemicals such as potassium permanganate and hydrogen peroxide. Odors become intensified when manure is wetted or when the waste is accumulated in a lagoon. Sulfurous compounds in the manure in the form of sulfides and/or mercaptans are major problems. These odorous substances can be controlled by the use of chemical oxidants such as those noted. Hydrogen peroxide at a weight rate of from 100 to 175 parts per million (ppm) may be required on a mass basis per mass of slurry treated. Permanganate concentrations of 1 to 2% at a pH of 8 to 9.5 are optimum for the proper oxidation reactions.

It is necessary to adequately protect livestock and other animals from exposure to these chemicals. Odors from animal/plant processing such as smokehouses, rendering and curing can be controlled by collecting the gases and passing them through a wet scrubber (absorber) or an adsorber. Either of these techniques can reduce the odor levels (usually expressed as odor units, o.u.) to within acceptable/allowable values. The same chemicals that have already been mentioned can be used in the wet absorbers. Oxidizing acids and bases can also be used in this type of control system. Costs of these procedures may range from $50 to $200 per acre per application, depending on the extent of the application.

Table 2.11 Dirt Road Control Techniques

Type of Material	Degree of Control, %	Estimated Cost, $/mi/application	Service Time, mo
Waste Oil[a]	70–80	10–50	2–3
Paper Mill Waste (lignens)	70–90	30–50	2–3
Chemical Encrusting Agents (glycols, etc.)	60–80	200–500	1–3
Clay and Chemical Binders ($CaSO_4$)	60–80	100–400	6–12

[a]Use may be prohibited.

Roads. Dust from driving on the farm roads can be excessive in dry periods under high-frequency traffic conditions. Fugitive dust from unpaved roads may average about 75 lb per vehicle mile of travel, as noted in Section 1.6.2.1. Paved roads and parking areas around farming operations can even contribute toward particulate emissions as depicted in Figure 1.7. These may be minor in quantity but significant because of relative location to working/living areas. Dust on unpaved farm roads can be controlled by wetting, by the use of binders or by application of gravel. Wetting is a temporary solution that presents a continuous operating problem, and that is why most people prefer the use of gravel or some type of binder such as those given in Table 2.11.

Many areas now restrict the use of waste oil because the runoff causes a water pollution problem. The best and most permanent solution is to pave the dirt road. This is also the most expensive, at $10,000 or more per mile. Surface treatment with chemical stabilizer could cost $2,000 or more per mile for 50% effectiveness in dust reduction. Application of a gravel surface could provide a 50% dust reduction [19].

Boilers. Emissions from boilers are well documented and for good facilities depend mainly on fuel type. Examples of emissions are given in Table 2.12 [20]. If required, particles can be controlled at over 99% efficiency by an electrostatic precipitator, wet scrubbers or filters. SO_x can be reduced by use of low sulfur fuels or it can be controlled at over 90% by wet scrubbers and at over 80% by dry scrubbers. The other pollutants are controlled by burner adjustments and modifications.

Other Combustion Sources. Open fires can release an average of 150 lb of particulate emissions per ton of combusted material, depending on the type of fuel [21]. This is small in comparison to the total emissions,

72 FUGITIVE EMISSIONS AND CONTROLS

Table 2.12 Small Stoker Emissions [20]

Type Boiler	Fuel	Emission Units	Average Uncontrolled Emissions[a]				
			Particulates	SO_x	CO	HC	NO_x
Commercial/ Domestic	Bituminous coal	lb/ton	2A	38S	10	3	6
	Anthracite	lb/ton	2A	38S	10	0.2	10
	Fuel oil	lb/10^3 gal	10	142S	5	3	12
	Natural gas	lb/10^6 ft^3	10	0.6	20	8	100
	LPG	lb/10^3 gal	1.9	0.09S	2	0.8	10
	Wood/bark	lb/ton	30	1.5	50	60	10
	Lignite	lb/ton	3A	30S	2	1	6
Hand-Fired Units	Bituminous coal	lb/ton	20	38S	90	20	3
	Anthracite	lb/ton	10	38S	90	2.5	3

[a] A = wt % ash in fuel, S = % sulfur in fuel.

which are 90% CO_2 and water vapor. The organic content of the released particulate matter is high, being about 10 times greater than that of normal atmospheric particulates [22]. CO emissions range up to 500 lb/ton for smoldering fires, and HC are 10 to 40 lb/ton. NO_x and SO_x are negligible. Control of these emissions is by good land management to prevent accumulation of combustible wastes and to control fires by use of prescription burning.

Wind Erosion of Exposed Areas. For completeness, it should be noted that dust emissions from open land can be controlled by use of three basic techniques in addition to a vegetation cover. These methods and costs, as adopted from open area emissions in the steel industry, are given in Table 2.13. Uncontrolled emissions amount to 1.4 tons/ha/yr.

2.3.2 Asphalt Concrete Plants

Fugitive dust consists of emissions from storage piles, transfer and conveyance operations, and screening. Figures 2.9a and b are schematics of a batch and a continuous plant. In addition to the fugitive emission sources listed in Table 2.14, the rotary dryer is a point source of particulate emissions.

The combined benefits from water sprays plus aggregate pile orientation could be:

$$[1 - (1 - 0.5)(1 - 0.5)] \text{ to } 1 - (1 - 0.75)(1 - 0.70) = 75 \text{ to } 93\%$$

CONTROL OF EMISSIONS 73

Table 2.13 Control of Wind Erosion of Exposed Areas [23]

Control Method	Cost in 1978 dollars, $/ha	Approximate Control Efficiency, %
Watering	2000	50
Chemical Stabilizers	2000 and up	70
Windbreaks	55–550	30

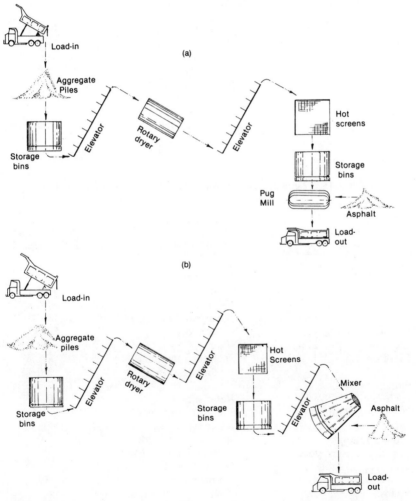

Figure 2.9 Asphalt concrete plants. (a) Batch plant. (b) Continuous plant.

74 FUGITIVE EMISSIONS AND CONTROLS

Table 2.14 Typical Asphalt Plant Fugitive Emissions

Source	Control Description	Approximate Efficiency, %
Load-in, aggregate	Water sprays	50–75
Aggregate storage piles	Wind screens	60–80
	Water sprays	50–75
	Pile orientation	50–70
Load aggregate to bins	Water sprays	50–75
Elevators (each), hot and/or cold	Water sprays	35
Screening aggregate	Enclosure ducted to dryer filter	90
Load-out	Enclosure ducted to dryer filter	90
Roads	Water spray	50
	Chemical stabilization	50
	Apply gravel	50
Open area	Vegetative cover	65
	Chemical stabilization	50

Table 2.15 Typical Cement Manufacturing Fugitive Emissions

Source	Control Description	Approximate Efficiency, %
Load-in, limestone and coal	Water sprays	50–75
Aggregate storage piles	Wind screens	60–80
	Water sprays	50–75
	Pile orientation	50–70
Crushers and grinders	Water sprays	50–75
Elevators (before kiln)	Water sprays	90
Elevators (after kiln)	Enclosure and duct to kiln baghouse	90
Storage bins/silos (before kiln)	Water sprays	50–75
Storage bins/silos (after kiln)	Enclosure and duct to kiln baghouse	90
Clinker fallout	Install clinker ladder	80
Transfer to conveyors	Hood and duct to filter	90
Product packaging	Enclosure and duct to kiln baghouse	50
Roads	Chemical stabilization Pave	85–90

2.3.3 Cement Manufacturing

A typical portland cement plant has numerous sources of fugitive emissions as it processes limestone to the cement product by the use of coal. A PEDCo study [7] lists 22 sources. In addition, the grinders and kilns are point sources of particulate emissions. Table 2.15 gives these fugitive sources as groups and shows control strategies that can be considered.

2.3.4 Foundries

All foundries use similar operations to convert the raw materials (metal ingots or scrap, chemicals and coke) to a metal product (aluminum, iron, steel, zinc, copper, brass or lead). These raw materials, plus sand and binders for the molds, are stored in piles. Typical foundry fugitive emissions could come from about 20 different sources. There are also about 7 point sources for particulate emissions. Table 2.16 lists fugitive emissions and possible controls.

2.3.5 Grain Handling

Grain from farms is transported to local elevators usually located within a 20-mile distance. These facilities process on the average each year about 15 times their storage capacity. The grains may be cleaned, dried and treated before storage or further shipment. The screening cleaners and dryers are point sources and are usually ducted to a baghouse. The major fugitive sources and possible controls at grain elevators are listed in Table 2.17.

2.3.6 Lime Plants

Lime plants use coal to heat limestone so that carbon dioxide is removed and lime remains. Approximately 60% of the emissions come from the secondary stone crushing and screening operations. The rotary lime kilns and coolers are point sources and usually exhaust through a baghouse. There are about 14 different fugitive sources in a typical plant. Table 2.18 lists some of these source groups and controls to consider.

76 FUGITIVE EMISSIONS AND CONTROLS

Table 2.16 Typical Foundry Fugitive Emissions

Source	Control Description	Approximate Efficiency, %
Load-in, raw materials	Water sprays	50–75
Aggregate storage piles	Wind screens	60–80
	Pile orientation	50–70
Roads	Chemical stabilization	50
	Pave	85
Open areas	Chemical stabilization	50
	Vegetative cover	65
Cupola, charging	Side draft hooding at charger discharge with exhaust to fabric filter	90
Cupola, tapping	Hood about tap through with exhaust to fabric filter	90
Cupola, regular operation	Improve monitoring and maintenance	99
Cupola, burndown (cupola shutdown)	Close charge port and make exhaust gas changes	99
Desulfurizing	Hood and duct to filter	90
Ductible Iron Treating	Improve maintenance	99
Fabric filter waster discharge & handling	Install screw conveyor to covered trucks	90
Plant solid waste dump	Spray and apply proper landfill practice	50

Table 2.17 Typical Grain Handling Fugitive Emissions

Source	Control Description	Approximate Efficiency, %
Load-in	Enclose and duct to baghouse	90
Transfer and conveyance	Enclose and duct to baghouse	90
Vents, all	Duct to baghouse	98
Load-out	Enclose and duct to baghouse	90
Roads	Pave	85
Open areas	Water sprays	50
	Pave	90

Table 2.18 Typical Lime Plant Fugitive Emissions

Source	Control Description	Approximate Efficiency, %
Load-in, limestone and coal	Water sprays	50–75
	Enclose	50
Aggregate storage piles	Water sprays	50–75
	Wind screens	60–80
	Orientation	50–70
Roads	Chemical stabilization	50
	Pave	85
Transfer to crusher	Water sprays	50–75
Crushing/grinding (before kiln)	Water sprays	50–75
Crushing/grinding (after kiln)	Hood and duct to filter	90
Screening (before kiln)	Water sprays	50–75
Screening (after kiln)	Hood and duct	90
Product storage	Vent to filter	98
Load-out	Vent to filter	90
Open areas	Vegetation	65

2.3.7 Mining Operations

The major mining operations include coal, copper and rock industries. Stone quarrying operations and phosphate rock mining are the main rock industry sources. PEDCo studies [24, 25] were made to estimate emissions from various mining operations. These results are summarized in Table 2.19. These values are appropriate and depend somewhat on the various industries. In general, copper has fewer overburden removal emissions but more storage emissions. Coal mining has less storage emissions and rock quarrying has more truck dumping emissions. These emissions will vary, as discussed in Chapter 1.

It appears that the emissions from the various coal mining sources are all quite uniform in size and size distribution and there does not appear to be a reduction in average particle size in the air with distance from source [25]. The mass mean diameter ranges from 18 to 23 μ and the standard geometric deviation is about 2.0. The dust is bimodal, with the larger portion equal to about 20% by mass >35 μ.

Table 2.20 lists the 11 mining operations emission sources and suggests typical control techniques.

78 FUGITIVE EMISSIONS AND CONTROLS

Table 2.19 Fugitive Emissions Ranges for Mining Operations [24,25]

Operation	Approximate Emission Range	Significance of Emissions by Industry[a]			
		Coal	Copper	Rock	Phosphate
Overburden removal	0.0008–0.45 lb/ton ore	5	3	3	5
	0.004–0.061 lb/ton overburden	3	5	5	0
Blasting	0.001–0.16 lb/ton ore	4	4	4	1
Shovel/truck loading	0.002–0.014 lb/ton ore	5	5	5	1
Haul roads	3.3–17.0 lb/vehicle mile traveled	3	5	5	1
Truck dumping	0.005–0.027 lb/ton ore	3	3	5	0
Crushing	≲0.7 lb/ton ore	2	2	2	3
Transfer and conveyance	≲0.2 lb/ton ore	1	1	1	1
Cleaning	negligible	3	3	5	5
Aggregate storage piles	0.0235–0.42 lb/ton ore 3.5–13.2 lb/(acre-day)	3	5	1	3
Waste disposal	≲14.4 ton/(acre-yr)				
Reclamation	(due to wind erosion)	4	1	3	4

[a] 5 = major source, 0 = negligible.

2.3.7.1 Limestone Crushing

Limestone crushing is presented as a special case because this operation is so significant in a number of industries (e.g., rock industry, power plants, construction and many metal producing industries). The National Crushed Stone Association (NCSA) conducted studies [26] on crushing operations to evaluate the effect of wet suppression on emissions of various sizes. These findings are given in Table 2.21. The efficiencies reported are compared to uncontrolled emissions. Note that the wet spray controls are in general more effective when larger particles are crushed (secondary crusher), but for each device the sprays are more efficient in reducing the finer dust material. Uncontrolled opacities are 91 and 92% for the secondary and tertiary crushers respectively.

2.3.8 Ore Handling Facilities

Ore handling facilities are a special category in themselves. They receive processed ore and simply transfer and store the material. Taconite

Table 2.20 Typical Mining Operations Fugitive Emissions

Source	Control Description	Approximate Efficiency, %
Overburden removal	Large buckets and reduced fall distance	30
Blasting	Optimum charge size	20
	Charge covers	30
	Schedule during low wind speeds	45
Shovel/truck loading	Water sprays	50–75
	Reduced fall distance	25
Haul roads	Water sprays	50–75
	Chemical stabilization	80
	Gravel	50
Truck dumping	Water sprays	50–75
	Enclosure	50
Crushing	Water sprays	50–75
	Enclosure	50
Transfer and conveyance	Water sprays	35
	Charged fogs	80
	Enclose, with spray	90
Cleaning	Water sprays	50–75
	Enclosure	50
Aggregate storage piles	Wind screens	60–80
	Pile orientation	50–70
	Leading slope angle	35
Waste disposal	Chemical stabilization	80
	Water sprays	50–75
Reclamation	Soil stabilization	80
	Vegetation	65

Table 2.21 Fugitive Emissions of Limestone from Crushers Using Wet Suppression Controls

	Controlled Emissions of	
	$<10\mu m$	$<50\mu m$
Emission Factor, lb/ton		
Secondary crusher	0.0048	0.0150
Tertiary crusher	0.0011	0.0016
Control Efficiencies, %		
Secondary crusher	92	83
Tertiary crusher	81	77
Opacity, %		
Secondary crusher	8	
Tertiary crusher	13	

80 FUGITIVE EMISSIONS AND CONTROLS

ore handling facilities are a significant example of this operation. Ore in the form of pellets is received in ore cars by rail. These cars are smaller than standard open cars because of the higher ore density. The ore is unloaded and transferred into ships, when available, or onto huge storage piles for later transport. For this reason some piles are relatively inactive. Table 2.22 lists typical fugitive sources and controls. Some load-in operations are point sources and are frequently ducted to a baghouse.

2.3.8.1 Case History

A taconite handling facility is used as an example case history to show how the emissions can be controlled by use of some intermediate techniques, as suggested in Table 2.22, and by long-range modifications and improvements. Note that control values for a specific operation can differ from those of Table 2.22 depending on the combination of techniques applied, the quantities, extent and frequency of treatment and the material being controlled.

Table 2.22 Typical Ore Handling Fugitive Emissions

Source	Control Description	Approximate Efficiency, %
Load-in, rail cars	Water sprays	50–75
	Enclosure	50
Transfer points	Wetting agent sprays	60–80
	Enclosure	50
	Reduced fall distance	25
Conveyance	Enclosure	50
	Wetting agent sprays	60–80
Roads	Water sprays	50–75
	Chemical stabilization	50
Aggregate storage piles, active	Pave	85
	Wind screens	60
	Pile orientation	50–70
	Leading slope angle	35
	Water sprays	50–75
Aggregate storage piles, inactive	Pile orientation	50–70
	Leading slope angle	35
	Encrusting	95
Load-out	Water sprays	50–75
	Enclosure	50
Open areas	Chemical stabilization	80
	Vegetation	65

Introduction. A transportation company operates an iron ore rail-to-ship bulk handling and storage facility. Almost all of the iron ore handled at the ore dock is in the form of taconite pellets. These pellets are the size of large marbles. The transportation company has operated a rail-to-ship transfer facility in this location for more than 100 years. The company presently plans to improve and upgrade the ore dock in order to increase tonnage throughput by 36% over present figures. To achieve this goal, the following components will be added to the existing system:

1. a new automatic three-car rotary dumper to replace the existing car dumper, which will only be utilized as a backup;
2. an improved crawler-mounted bucket wheel reclaimer to complement the original reclaimer;
3. capacity improvements to several existing conveyor belts; and
4. consolidation and enclosure of transfer stations, as well as selected conveyor belts.

Existing Facility. Iron ore is delivered by switch engines to the automatic rotary car dumper in 51 car cuts, with each car carrying 70 tons of pelletized taconite ore. The car dumper can empty three cars in a single rotation. As a set of three loaded cars is moved into the north end of the dumper, an automatic positioner arm simultaneously shoves a set of three empty cars out the south end, where they couple with other empty cars and are held until a sufficient number has accumulated, at which time they are returned to the mines for loading.

The rotary car dumper is enclosed in a building containing a baghouse collection system as well as a suppressant spray system at strategic locations, all of which prevents fugitive emissions from escaping into the ambient environment during dumping operations. The collected dust from the baghouse is stored in specially designed covered hopper cars which, when full, are returned to the pelletizing plants for reprocessing.

The ore is carried from the dumper by a conveyor system through a transfer station designed to allow direct flow to ship or stockpile. During winter months when shipping movements are suspended, as well as during periods when ships are not being loaded, the ore is sent to storage stockpiles. Approximately 80% of incoming ore is stockpiled initially and the remaining 20% is loaded directly to ships. All of the ore is moved on a system of steel cable-reinforced rubber conveyor belts ranging in width from 5 to 8 ft. There is a total of approximately 4 mi of belting in the system.

During stockpiling, the ore pellets pass through the transfer point to the stacker for placement on the piles. The stacker is a self-propelled, rail-mounted unit designed to parallel the length of a 3100-ft-long storage area on a raised earth berm. From this rail bed, the stacker has sufficient lateral mobility to distribute several grades of taconite pellets onto one of six stockpiles. To retrieve ore from the stockpile for shiploading, a self-propelled crawler-mounted reclaimer is employed. This unit has the capability to dig into the stockpiled ore, at a rate of 5600 tons/hr, by means of a large rotating wheel equipped with scoop buckets. On every revolution, the buckets deposit the ore onto an internal conveyor belt which transfers the reclaimed ore to the primary conveyor system for shiploading.

When the reclaimer is working the outer limits of the stockpile, the distance exceeds the length of the discharge boom, thus preventing the unit from discharging ore directly onto the primary conveyor belt. To accommodate this phase of reclamation, a mobile transfer conveyor, called a bandwagon, is used to connect the reclaimer to the conveyor system.

The ore, enroute to the ship, is directed from the primary reclamation conveyor back through the transfer point to a structure referred to as a surge bin. The bin, similar to a silo, contains enough volume to store ore in variable amounts during shiploading. This permits reclaiming operations to continue uninterrupted while the shiploader temporarily ceases loading to switch hatches on the ship.

Ore released from the surge bin is transferred to a series of conveyors that direct material to the shiploader. This unit, also rail-mounted, is similar in operation to the stacker. It can travel the length of the dock in forward or reverse direction and allows the loading conveyor boom to service ships on either side of the dock.

The dock can accommodate ships with capacities in the range of 6000 to 60,000 tons. In comparative terms, a ship with a 6000-ton capacity can accommodate 85 cars of ore while a 60,000-ton ship requires the contents of 860 cars to reach capacity.

Requirements. Presently, the ore dock under consideration is in a secondary nonattainment area since it does not meet air quality standards for suspended particulates, i.e., 150 $\mu g/m^3$ maximum 24-hr concentration not to be exceeded more than once per year, or 60 $\mu g/m^3$ as the annual geometric mean. Based on air quality data for suspended particulates accumulated and analyzed for the area for 1976 and 1977 by the state, it was determined that the average maximum 24-hr value was exceeded a sufficient number of times to designate it as nonattainment.

In discussions with the local environmental regulatory agency regarding the ore dock modifications, it was required that the company utilize

CONTROL OF EMISSIONS   83

offsets. The emission offset policy takes effect in an area of nonattainment. It requires that: new emissions must be controlled to the greatest degree possible, must be more than offset by emission reductions from sources already existing in the area, and the offset procedure must lead to progress toward achievement of ambient air quality standards. An offset factor of 150% was assigned to the facility that stipulated that for every pound of particulates produced by a new source, there must be a corresponding reduction of 1.5 lb from the remainder of the facility.

A second requirement imposed by the state subjected the company to the lowest achievable emission rate (LAER) standards. LAER can be defined as the best emission control, regardless of cost, which has ever been implemented to control emissions from a facility of this type.

LAER Controls. An approach to upgrade the air quality in the area, under LAER, includes the following controls:

1. revision of existing baghouse dumper ductwork to improve collection efficiency,
2. installation of a baghouse collection system in a new dumper,
3. enclosure of all transfer locations,
4. enclosure of all conveyor systems which would not impair operation of equipment,
5. refinement and addition to existing wet spray suppressant system in dumpers and at transfer locations,
6. spraying of active stockpiles at frequent intervals and encrusting of inactive piles,
7. road and yard spraying to reduce fugitive dust, and
8. installation of water treatment system to remove solids from general cleaning operations and spraying operations.

Emission Estimates. It is difficult to obtain reliable emission estimates for the fugitive emissions from bulk handling facilities, and taconite terminals are no exception. Emission source inventories must be based on good engineering judgment and data from available literature such as those given in Chapter 1 and Section 2.1. Due to the disparity of published data, assistance was obtained from the local agency for this specific facility. The proposed uncontrolled emission factors and emission estimates are listed in Table 2.23. Using these factors this existing terminal could emit up to 1659 tons/yr of particulates.

It is important to notice that the emissions are based on an emission factor (lb/hr of emissions per tons/hr handled) and a use factor (proportion of total throughput of the facility that is handled by each

84 FUGITIVE EMISSIONS AND CONTROLS

Table 2.23 Summary of Uncontrolled Emissions for Example Taconite Terminal

Use Factor	Transfer Point[a]	Description of Transfer Point	Existing Control	Current Emissions, tons/yr
1.0	Existing car dumper	Railroad car unloading bldg	Drench/spray baghouse	22.2
0.80	T1P	C2 to C3	Spray	5.4
0.20	T2P	C2 to C4	Spray	1.4
0.80	T3P	C3 to C3A	Previously sprayed	21.6
0.80	T4P	C3A to pile	Previously sprayed	24.5
0.80	T5P	Pile to reclaimer boom	None	198.0
0.80	T6P	Reclaimer to discharge boom	None	144.2
0.80	T7P	Discharge boom to bandwagon	None	144.2
0.80	T8P	Bandwagon to C8 hopper	None	72.1
0.80	T9P	C8 to C4	None	144.2
1.0	T10P	C4 to Surge bin	Enclosed	180.0
1.0	T11P	Surge bin to C5	None	90.0
1.0	T12P	C5 to C6	None	180.2
1.0	T13P	C6 to C6A	None	180.2
1.0	T14P	C6A to ship	None	120.0
	Wind Erosion	Storage pile wind erosion	None	127.0
	Roads	Vehicular traffic	None	4.0
Total				1659.2

[a]See Figure 2.10 for transfer locations.

piece of equipment). The ore handling rate for this facility is 12×10^6 tons/yr. The storage rate is 2.5×10^6 tons.

The transfer point locations are shown in Figure 2.10.

Expanded Facility. This terminal is to be upgraded and expanded and will therefore require LAER controls. It is important to estimate what the acceptable LAER controls are and what the modified facility would emit after application of these LAER controls. Table 2.24 summarizes the controls for each unit operation and gives the estimated emissions after control. Refer to Figure 2.10 for source locations. Note that now total emissions are 470 ton/yr, which is 71% less than the previous emission total.

CONTROL OF EMISSIONS 85

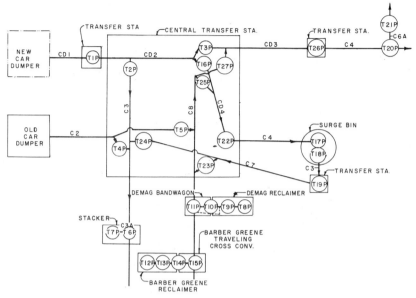

Figure 2.10 Example taconite terminal transfer points.

Most of the control is based on a concept of keeping the taconite wetted, with the exception of where the pellets enter the terminal. The dump stations are equipped with baghouses. The baghouses are rated at 99.8% efficiency for the baghouse per se, but the overall controls including dust collection provide an emission reduction of only 60%.

The air pollution control system for the existing taconite rail car dumping station was not designed correctly in that the dust collection system is almost totally inadequate. When the unit was initially designed, the consultant felt that most of the dust would be generated on the side of rotation of the rail car, and placed most of the collection piles along this side of the dump house at the grizzley level. Unfortunately, the falling pellets form a seal, and the dust comes up on the other side where only a minimum of collection pipes was installed.

Dust generated by the dumping of railroad ore cars filled with pelletized taconite ore is pulled from the dumping containment area by an induced-draft fan. The dust can then be treated as a particulate emission exhaust from a process operation. Air flow in the transfer ducts is turbulent at the 67 ft/sec velocity, which helps reduce settling of this heavy iron ore dust mixture, which has a particle density of about 5.5 g/cm^3.

Particle measurements were made with a standard EPA Method 5 test train. Dust loadings were determined using the Method 17 (in stack

Table 2.24 Summary of LAER Controls and Emissions for the Taconite Facility

Transfer Point/Source	Description of Transfer Point/Source	Proposed LAER Control Technique	Control Efficiency	Estimated (tons/yr) Annual Emissions
New car dumper	New railroad car unloading building	Sprays/scrubber	99.4	176.00
Existing car dumper	Existing railroad car unloading building	Sprays/baghouse	99.9	4.16
T1P	CD1 to CD2	Enclosed/spray	95.0	8.36
T2P	CD2 to C3	Enclosed/spray	95.0	6.69
T3P	CD2 to CD3	Enclosed/spray	95.0	1.48
T4P	C2 to C3	Enclosed/spray	95.0	1.18
T5P	C2 to C8	Enclosed/spray	95.0	0.30
T6P	C3 to C3A	Previously treated	90.0	15.70
T7P	C3A to Pile	Previously treated	90.0	22.30
T8P	Pile to DEMAG Reclaimer boom	Previously treated	75.0	22.60
T9P	Reclaimer boom to discharge boom	Previously treated	75.0	19.70
T10P	Discharge boom to bandwagon	Previously treated	75.0	19.70
T11P	Bandwagon to C8	Previously treated	75.0	8.90
T12P	Pile to BARBER-GREENE Reclaimer	Previously treated	75.0	22.60
T13P	Reclaimer to discharge Boom	Previously treated	75.0	19.70
T14P	Discharge boom to traveling conveyor	Previously treated	75.0	19.70
T15P	Traveling conveyor to C8	Previously treated	50.0	8.90
T16P	CD2 to CD4	Enclosed/spray	95.0	0.20
T17P	C4 to surge bin	Enclosed/previously treated	95.0	0.96
T18P	Surge bin to C5	Partially enclosed/previously treated	95.0	0.96
T19P	C5 to C7	Enclosed/previously treated	95.0	0.96
T20P	C6 to C6A	Previously treated	50.0	29.50
T21P	C6A to ship	Previously treated	50.0	19.70
T22P	CD4 to C4	Enclosed/spray	95.0	0.98
T23P	C7 to C8	Enclosed/previously treated	95.0	0.96
T24P	C7 to C3	Enclosed/previously treated	85.0	0.00
T25P	C8 to CD4	Enclosed/spray	95.0	0.78
T26P	CD3 to C6	Enclosed/spray	95.0	9.80
T27P	C8 to CD3	Enclosed/spray	95.0	8.40
Storage Pile	Wind erosion	Sprays	90.0	17.40
Road	Vehicular traffic	Chemical binder	60.0	1.60
Total				470.17

thimble) adaptation. Dust concentrations vary both with time and with type of ore being dumped. Maximum concentrations occur just as the cars are dumped and decrease with time until the next cars are positioned and dumped. For a normal 3-min dump cycle, the maximum dusting occurs over a 35-sec period. A time-integrated dust concentration for normal dumping cycles averages about 2.28 gr/acf. The dust appears reddish-black and, although dry, has a slippery feel.

Taconite ore dust is bimodal in distribution as shown in Figure 2.11. The larger portion, which is black, represents 77% of the dust by mass and is crystalline in structure. The smaller 23% is red and noncrystalline, indicating that this would be more Fe_2O_3. A small amount of silica crystals was observed in the smaller-sized fraction of the dust.

The mass mean diameter of the dust is 10 μ. The standard geometric deviation is 2.6 for the larger and 9.2 for the smaller-sized fractions respectively. To increase the dust collection efficiency within the existing car dumper it is necessary to:

- add small hoods at the ends of the car dumper on the side of initial rotation, and connect these to two small baghouses
- add large hoods next to the grizzley on the far side of the car dumper at the grizzley level.

2.3.9 Coal-Fired Power Plants

Many fugitive emission sources at coal-fired power plants are the same as those from coal mining, and if limestone/lime flue gas scrubbers are used these dust sources are similar to rock industry emissions. Many conventional and novel techniques have been summarized by Currier and Neal [27]. Some of these and other data are presented in Table 2.25 for a typical coal-fired utility. Control costs are included where available and are in 1978 dollars. As always, many of these control suggestions can be duplicated in the various operations and are not always repeated in the suggested listing in the table. In addition, the stacks are obviously point sources of particulate emissions.

2.3.9.1 Case History

This example describes possible fugitive emissions from a large 2000-MW coal-fired utility. It also discusses impact significance of these *controlled* emissions. Note that included in the fugitive emission sources is the evaporation from spray cooling towers. Should these be present, they

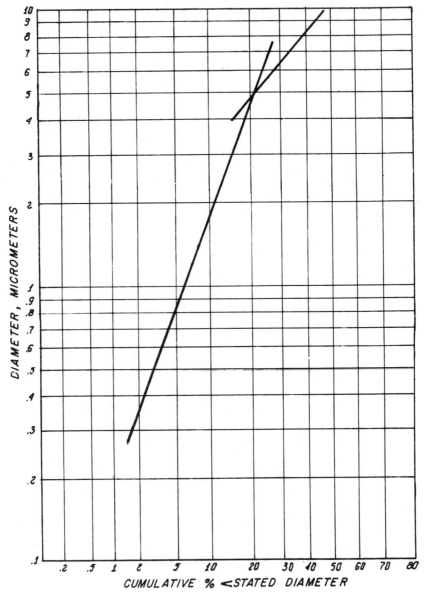

Figure 2.11 Taconite ore dust size and size distribution by mass.

will be sources of emission. A special note is made of possible hazardous emissions. A limestone flue gas scrubbing system is assumed in this example.

Table 2.25 Typical Coal-Fired Power Plant Fugitive Emissions

Source	Description	Control Approximate Efficiency, %	Cost
Load-in, coal	Water sprays	50–75	
	Enclose	50	
	Negative pressure, enclosure and duct to filter	98	
Storage piles	Cap with large coal Enclose	95	$72–172/ton
	Chemical sprays	60–90	
	Water sprays	50–75	
	Wind screens	60–80	
	Pile orientation	50–70	
	Leading slope angle	35	
Transfer, coal	Enclose	50	
	Water sprays	50–75	
	Chutes	75	
	Reduced fall distance	25	
	Chemical sprays	85	
Transfer, conveyor	Enclose	50	
	Hood and duct to filter	98	$16,000 per station
	Water sprays	50–75	
	Foam sprays	80	
	Chemical sprays	60–90	
Roads	Water sprays	50}	$11,000/ truck + $21,500/yr
	Chemical suppression spray	85	
Open areas	Chemical stabilization	80	
	Vegetation	65	
Load-in/out, limestone	Water sprays	50	
	Enclose	50	$43–102/ton
	Chutes	75	$5.2–8.3/ (ton-hr)
	Reduced fall distance	25	
Waste disposal	Chemical stabilization	70	>$2000/ha
	Windbreaks	30	$55–550/ha
	Vegetation	65	

Facility and Assumption. This is a theoretical facility in the Northwestern United States and certain assumptions on the general arrangement of the power station and the types of equipment for coal handling are made. Figure 2.12 is a sketch of the general arrangement of the plant

90 FUGITIVE EMISSIONS AND CONTROLS

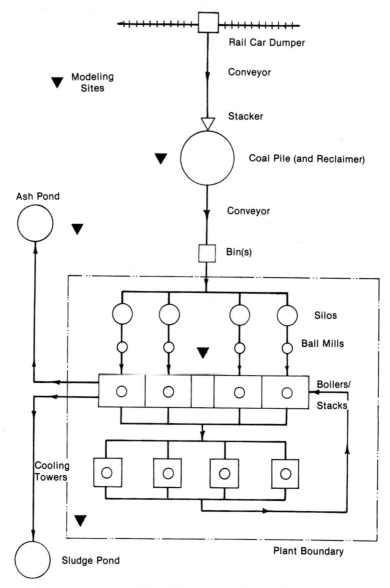

Figure 2.12 Hypothetical 2000-MW generating station plant arrangement including a simplified flow diagram.

indicating the size and location of coal piles, conveyors, cooling towers and train unloading facilities. Table 2.26 includes a list of specific control systems considered for the fugitive sources of emissions, and the

Table 2.26 Inventory of Fugitive Emissions from an Example 2000-MW Western Generating Station at 75% Capacity Factor

Emission Source	Basis of Emission Calculation	Uncontrolled Emission Factor (lb/ton)	Uncontrolled Emission Rate (lb/hr)
Rail car unloading	200 car dumps/day	0.007	6
Conveyors to pile	0.5 mi	0.2	170
Stacker	20,000 tons/day	0.037	30
Pile	40 ft high, 30.6 acres (90-day supply)	6.4 lb/acre/hr	200
Reclaimer	20,000 tons/day	0.2	170
Reclaimer conveyor	0.5 mi	0.2	170
Bins	Single	0.0002	nil
Silos	Four (units)	0.0002	nil
Cooling towers	Four units at 0.005% of total recirculated water	5.5 lb/hr per tower	22
Roads	Estimation 300 mi/day	2.4 lb/vehicle/mi	450
Ponds, total	100 acres (dry)	1.2 lb/acre/hr	120
Limestone unloading	1.6 lb/100 lb coal	0.007	nil
Limestone storage	90-day supply	6.4 lb/hr	4
Limestone crushing and transfer	0.5 mi	0.2	3
Emergency equipment	50 hr/yr	0.2	nil
Emergency boiler	Oil-fired/40 days	1	1
Total uncontrolled emissions			1346

specific estimated emissions from these sources. A 90-day coal and chemical storage pile quantity is assumed. A 1.1 limestone supplied to SO_2 removed stoichiometry is also assumed. The coal pile storage area is large and covers about 40 acres, with the piles stacked 30 ft high.

92 FUGITIVE EMISSIONS AND CONTROLS

It is possible that two 100-car-unit trains could be dumped each day at full load operation of the utility. This would amount to a transfer of 20,000 tons/day of coal. Actual transfer would occur over a 6- to 8-hr period of time. The values shown in Table 2.25 are for uniform, continuous unloading, so these emissions are lower than actual unloading period values.

Note that cooling tower drift is included as contributing toward the total fugitive particulate emissions. The towers, like the coal piles, release the emissions at a relatively low level compared with typical stack emissions, so they can be included with the other low level emissions. The water mist leaving a cooling tower contains solids, as shown in Table 2.27. Film rather than splash packing and low air velocities in the towers are used to help minimize salt losses.

The emissions from the first eight entries in Table 2.26 would be essentially coal dust. In addition, the emissions from the roads and ponds would contain considerable coal dust. This type of dust accounts for about 85% of all the fugitive dust noted. The composition of this dust would consist of the same components as the coal, including the radionuclides, which are shown in Table 2.30.

Fugitive Emissions Control. Various techniques can be used to control fugitive emissions. Some of these are summarized in Table 2.28 for three categories. Table 2.28 source numbers follow the emission source order of Table 2.26. The lesser degree of control is listed as RACT (reasonable available control technology), the more extensive control is listed as BACT (best available control technology) and the greatest control as LAER (lowest achievable emission rates). All three of these inter-

Table 2.27 Typical Cooling Tower Water Quality
(average concentration values in mg/l at 18 cycles/blowdown)

Calcium	364
Magnesium	81
Sodium	38
Potassium	20
Alkalinity or $CaCO_3$	275
Sulfate	950
Chloride	96
Fluoride	3
Ammonia and Nitrate as N	9
Phosphorous	2
Organic Carbon	29
Silica as SiO_2	92
Inhibitor	30

Table 2.28 Controlled Fugitive Emissions from a 2000-MW Western Generating Station at 75% Capacity Factor

Emission Source	RACT CONTROLS			BACT CONTROLS			LAER CONTROLS		
	Type	Eff., %	Emissions, lb/hr	Type	Eff., %	Emissions, lb/hr	Type	Eff., %	Emissions, lb/hr
1. Rail car unloading	Water sprays	35	4	Wetting agent, fog sprays	80	1	Enclosure and sprays	90	1
2. Conveyors to pile	Water sprays	35	111	Wetting agent, fog sprays	80	34	Enclosure and sprays	90	17
3. Stacker	Water sprays	35	20	Wetting agent, fog sprays	80	6	Enclosure and sprays	90	3
4. Storage piles	Water wetting, pile orientation, and slope angle	90	20	Wetting agents, wind screens and orientation	95	10	Encrusting	98	4
5. Reclaimer	Water sprays	35	111	Wetting agent, fog sprays	80	34	Enclosure and sprays	90	17
6. Reclaimer conveyor	Water sprays	35	111	Wetting agent, fog sprays	80	34	Enclosure and sprays	90	17
7. Bins	Enclosure	95	nil	Enclosure	95	nil	Enclosure	98	nil
8. Silos	Enclosure	95	nil	Enclosure	95	nil	Enclosure	98	nil
9. Cooling towers	Design		22	Design		22	Design	30	15
10. Roads	Water wetting and speed control	65	158	Wetting agents, speed control, gravel and stabilization	80	90	Paving and sweeping	85	68
11. Ponds, total	Wind screens	60	48	Stabilization and wind screens	95	6	Encrusting	98	2
12. Limestone unloading	Water sprays	35	nil	Wetting agent, fog sprays	80	nil	Enclosure and sprays	90	nil
13. Limestone storage	Water wetting, Pile orientation, and slope angle	90	1	Wetting agents, wind screens and orientation	95	nil	Encrusting	98	nil
14. Limestone Crushing and transfer	Water sprays	35	2	Wetting agent, fog sprays	80	1	Enclosure and sprays	90	nil
15. Emergency equipment	Design		nil	Design		nil	Design		nil
16. Emergency boiler	Design		1	Design		1	Design	50	nil
Total controlled emissions			609			239			144

CONTROL OF EMISSIONS 93

94 FUGITIVE EMISSIONS AND CONTROLS

pretations are the authors' and are as of this date. Descriptions of these control techniques, with supporting references, are detailed in Chapter 1 and the previous portions of this chapter.

The RACT controls could reduce overall fugitive emissions by 55%, BACT controls are 82% effective overall and LAER is 89% effective. Obviously, each of these improved control levels can be achieved only by the expenditure of additional funds.

Coal. An analysis of typical western coal is given in Table 2.29. The composition of ash is included on an ignited basis and shows what ash pond emissions would include. Included as "others" are such substances as antimony, asbestos, arsenic, barium, beryllium, boron, cadmium, cobalt, chromium, copper, manganese, mercury, nickel, selenium, vanadium and zinc.

All coal contains trace amounts of radioactive nuclides such as:

Uranium—238
Uranium—234

Table 2.29 Typical Western Coal and Ash Analysis

Substance	Parameter	Proximate Analysis, %	Utimate Analysis, %
Coal	Heating value, Btu/lb	9000	
	Moisture	23	23
	Volatile matter	30	
	Fixed carbon	37	
	Ash	10	10
	Carbon		51
	Hydrogen		3.3
	Nitrogen		1
	Chlorine		0.01
	Sulfur		0.7
	Oxygen		11
Ash	Silica		39
	Alumina		18
	Ferric oxide		6
	Titania		1
	Phosphorous pentoxide		1
	Lime		15
	Magnesia		3
	Sodium oxide		2
	Potassium oxide		1
	Sulfur trioxide		13
	Other		1

Lead—210
Polonium—210
Thorium—230
Radon—222
Radium—226

Concentrations of these elements in typical western coal may be approximately equal, or may vary depending on the coal seam. Radiation values for western coal are given in Table 2.30. The Pb-210 and Po-210 are volatile and are thus captured (and concentrated) in the fly ash and sludge. Uranium is concentrated in both bottom ash and fly ash. The radon is gaseous and is released with the stack gases.

Dispersion Models and Wind Data. Mountainous terrain is assumed for the location of the power station, and therefore the EPA VALLEY Model is used to estimate downwind ground level concentrations (GLC). The meteorological data used in the VALLEY Model calculations for the 3-hr winds are presented as a frequency distribution in Table 2.31. These data are restrictive, but are reported in the mountainous regions of the western United States and would apply for example in SO_2 GLC concentrations where a 3-hr prevention of significant deterioration (PSD) standard applies. EPA "default" data are used for the 24-hr meteorology that is applicable for the particulate GLCs presented in this chapter.

A 10-m plume height is used in these calculations for all ground level emissions sources that are actually less than 10 m high. The fugitive emissions sources were considered to originate at four locations throughout the area, as shown in Figure 2.12 by the marks, ▼. Appropriate coordinates for these are used as inputs to the EPA VALLEY Model.

All models are subject to certain imperfections and deficiencies. Yet certain models do consider factors that at least attempt to correct for conditions not considered in other models. It is recognized that Williams and others are in the process of developing high terrain models which may be more accurate, yet at this time the EPA accepts the VALLEY Model as being most applicable to mountainous terrain modeling. Data

Table 2.30 Radioactive Trace Elements in Typical Western Coals

Element	Range, pCi/g
U-234	0.13–0.24
U-238	0.12–0.24
Po-210	0.10–0.27

96 FUGITIVE EMISSIONS AND CONTROLS

bases are being obtained for attempted validation of the EPA VALLEY and CRSTER (for conventional terrain) models and for other comparable models such as the "Texas" and "Small Hill" models. It is not the intent of this example to attempt to validate the VALLEY Model, nor is it desired to extend the dispersion predictions to the 1 $\mu g/m^3$ level over hundreds of miles, as has been done by others. The VALLEY Model is used because it is the only high terrain model currently accepted by the EPA without validation [28].

Terrain. A schematic of the downwind terrain showing relative distances and elevations is given in Figure 2.13. This figure is not to scale, and shows only the specific locations corresponding to the wind directions used in Figure 2.14. Obviously, other wind directions and terrain conditions could be used. The ones used give the severest impacts.

Impact Analysis Results. Data are shown for GLCs resulting from the total fugitive emissions. The data are for all receptors for 24-hr wind persistence averaging times for the most restrictive meteorological conditions. The GLC data are calculated [29] and are given in Figure 2.14. For the four cases, i.e., for: (1) uncontrolled emissions, (2) RACT, (3) BACT, and (4) LAER, a screening series of model runs was made to establish the most restrictive 3-hr meteorological data—which turned out to be Case No. 4 as shown in Table 2.31 for the particular wind directions and terrain chosen for this study. An actual wind speed of

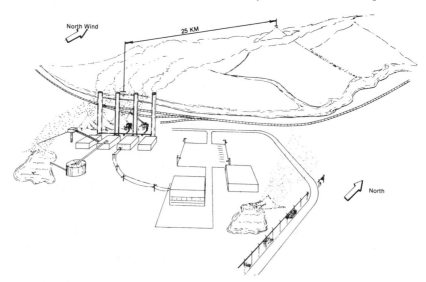

Figure 2.13 Panoramic sketch of an example generating station and local topography.

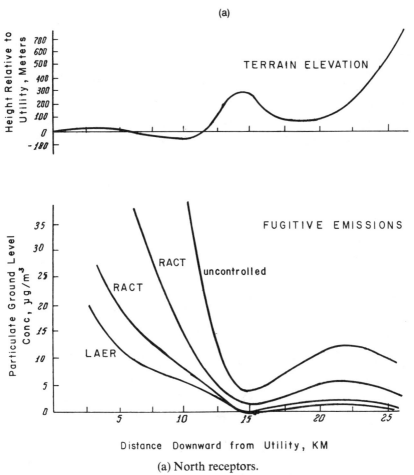

(a) North receptors.

Figure 2.14a 24-hr downwind particulate GLC due to north wind and EPA Star meteorological data for receptor locations is noted with respect to the 2000-MW utility.

2.6 m/sec and stability Class 6 would apply. These GLC data are not applicable to particulate PSD requirements so they are not included here. (The magnitudes of the 3-hr values are about 3.1 times larger than those of the 24-hr values.)

A maximum GLC (Figure 2.15) occurs at the NNE receptor Number I. The magnitude of this GLC is 210.5 µg/m³ for uncontrolled emissions, 95.2 µg/m³ for RACT controlled emissions, 37.4 µg/m³ for BACT and 22.5 µg/m³ for LAER. If NNE wind direction were used in the model this value would be higher. A profile of receptor heights is

98 FUGITIVE EMISSIONS AND CONTROLS

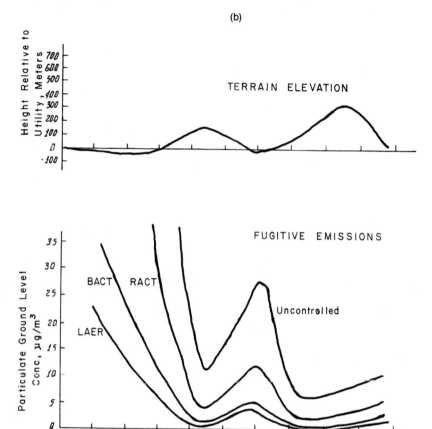

(b) North-northeast receptors.

Figure 2.14b 24-hr downwind particulate GLC due to north wind and EPA Star meteorological data for receptor locations as noted with respect to the 2000-MW utility.

shown above each part of Figure 2.14. Figure 2.14c shows that for relatively smooth terrain the downwind GLCs of fugitive dust decrease exponentially with distance. All GLCs decrease with deviation from source wind direction. Figure 2.14a indicates that a large-elevation receptor would block further significant transport of fugitive emissions, whereas Figure 2.14b suggests that a lesser receptor elevation could result in higher fugitive downwind GLCs at greater distances.

Effects of trace elements and radioactive nuclides can be estimated

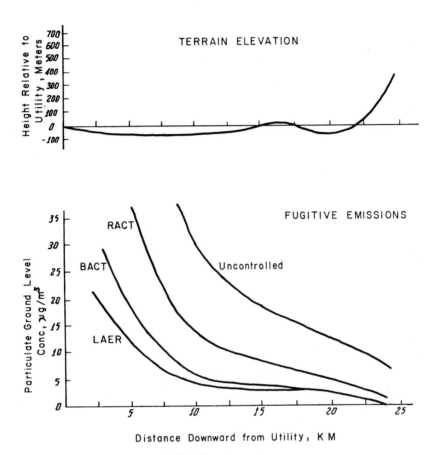

(c) Northeast receptors.

Figure 2.14c 24-hr downwind particulate GLC due to north wind and EPA Star meteorological data for receptor locations as noted with respect to the 2000-MW utility.

by multiplying the reported GLCs by a scaling fraction. This fraction would be the mass of substance in question divided by the mass of total fugitive emissions as shown in Table 2.26. For example, emissions of substances listed as "other" in the ash would amount to (0.01)(15%) or 0.1% of the coal dust plus up to about 1% of the pond dust. The radionuclide GLCs could be estimated in a similar manner. For example, radionuclide concentrations could be increased 10 times by separation of ash from the coal. Bottom and fly ash may be mixed and then dis-

(d)

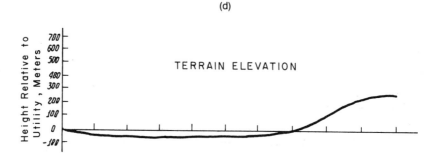

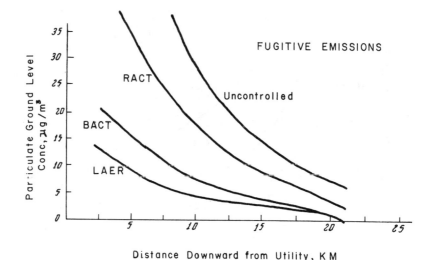

Distance Downward from Utility, KM

(d) North-northwest receptors.

Figure 2.14d 24-hr downwind particulate GLC due to north wind and EPA Star meteorological data for receptor locations as noted with respect to the 2000-MW utility.

posed of by ponding. Using an average radiation level for coal of 0.20 pCi/g, the GLC's radiation level could be about 0.31 pCi/g from fugitive emissions not considering decay. This is slightly higher than the natural radiation level of the earth's crust. Using similar suppositions, maximum GLCs predicted for the various hazardous pollutants are given in Table 2.32. All these values are less than any currently existing acceptable air concentration limit.

Summary. Downwind particulate GLCs from a hypothetical 2000-

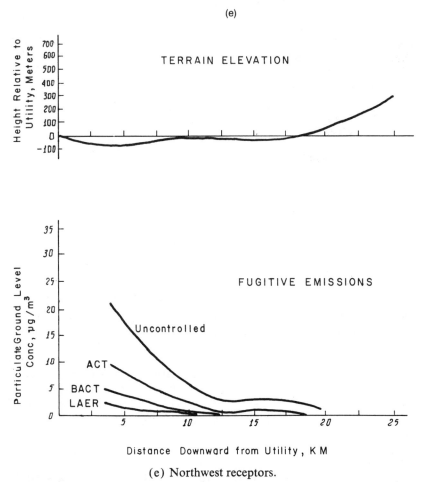

(e) Northwest receptors.

Figure 2.14e 24-hr downwind particulate GLC due to north wind and EPA Star meteorological data for receptor locations as noted with respect to the 2000-MW utility.

MW coal-burning utility depend significantly on the extent of control of fugitive emissions. In this example with very mountainous terrain and therefore extremely adverse dispersion capability, even the most controlled situation could result in the PSD increments being exceeded. If more normal downwind terrain exists, as for the example in Figure 2.14c, the BACT and LAER emissions are below Class II PSD limits at essentially all offsite locations, whereas the RACT exceeds the PSD limits at distances of less than about 7 kM. Uncontrolled emissions ex-

Table 2.31 Meteorology 3-hr Wind Matrix[a]

Case No.	Description	Wind Speed, m/sec	Wind Direction degrees	Stability Class
1	Low wind speed Unstable	1.0	206	1
		2.0	192	1
		2.2	203	1
2	High wind speed Unstable	8.9	185	4
		8.4	179	4
		10.1	183	4
3	Low wind speed Stable	0.7	181	6
		1.8	179	6
		1.8	187	6
4	Medium wind speed Stable	2.6	189	6[b]
		2.5	186	6
		2.8	182	

[a] Personal files.
[b] Case 4 is most restrictive.

ceed the PSD limits for distances up to about 12 kM. These comparisons show the significance of BACT and LAER controls. The particulate emissions may contain toxic substances which may be completely within

Table 2.32 Estimated 24-hr Maximum GLCs from Fugitive Dust Emissions at the example Western 2000-MW Coal-Fired Generating Station

Pollutant	Maximum Ground Level Concentration, $\mu g/m^3$			
	Uncontrolled	RACT	BACT	LAER
Antimony	1.6×10^{-3}	7.4×10^{-4}	2.9×10^{-4}	1.8×10^{-4}
Asbestos[a]				
Arsenic	7.8×10^{-3}	3.5×10^{-3}	1.4×10^{-3}	8.4×10^{-2}
Barium	16.0	7.2	2.8	1.7
Beryllium	2.7×10^{-2}	1.2×10^{-2}	4.6×10^{-3}	2.8×10^{-3}
Boron	1.0	0.46	0.18	0.11
Cadmium	4.5×10^{-2}	2.0×10^{-2}	7.9×10^{-3}	4.8×10^{-3}
Cobalt	3.3×10^{-2}	1.5×10^{-2}	5.8×10^{-3}	3.5×10^{-3}
Chromium	0.16	0.07	0.03	0.02
Copper	0.23	0.11	0.04	0.03
Manganese	0.70	0.32	0.13	0.08
Mercury	1.9×10^{-2}	8.5×10^{-3}	3.3×10^{-3}	2.0×10^{-3}
Nickel	0.08	0.04	0.01	0.01
Selenium	1.8×10^{-2}	8.1×10^{-2}	3.2×10^{-3}	1.9×10^{-3}
Vanadium	0.31	0.14	0.06	0.03
Zinc	0.55	0.25	0.10	0.06

[a] Essentially nonexistent.

safe and acceptable levels, but should be considered when determining total effects including background and all other sources.

In addition to showing the significance of various control levels, this example also shows the effects of downwind terrain on GLC air quality. These are:

1. Fugitive dust GLC decreases with downwind distance for relatively smooth terrain.
2. Fugitive dust GLC decreases as source wind direction deviates from source-receptor direction.
3. A large-elevation receptor (>200 m above source) may prevent further downwind transport of fugitive dust.
4. Substantial downwind elevations of <200 m may result in higher fugitive GLCs at greater distance.

Table 2.33 Typical Smelting Operations Fugitive Emissions

Source	Control Description	Approximate Efficiency, %
Load-in, ore and chemicals	Windbreaks	50
	Water sprays	50–75
	Enclosure	50
Material transfer	Water sprays	50–75
	Enclosure	50
Storage piles	Pile orientation	50–70
	Water sprays	30
	Wind screens	60
Metal/slag pouring	Enclose and duct to collector	90
	Charged fog sprays	90
Furnace dust	Fogging sprays	80
Casting	Duct ventilation to collector	90
	Fogging sprays	80
Mixing/preparing	Enclose	50
Crushing	Windbreaks	50
	Water sprays	50–75
	Chemical sprays	60–90
Fume leakage	Flogging sprays	80
	Charged fog	90
Roads	Water sprays	50–75
	Chemical stabilization	50
	Pave	90
Open areas	Chemical stabilization	80
	Vegetation	65

2.3.10 Nonferrous Smelting Operations

Smelters basically reduce ores to their metallic components. This includes aluminum, copper, lead, zinc, beryllium and other metal industries. The primary smelters work with the metallic oxide ores plus chemical fluxes and other agents, whereas the secondary smelters refine metal pigs, scrap and foundry rejects. This section deals with smelting in general without attempting to distinguish between the different types. This is reflected in the typical fugitive emissions listing of Table 2.33. Point sources of emissions include the smelting furnaces, roasters and sintering.

2.3.11 Woodworking Operations

Fugitive emissions at sawmills come from preparing and working with the wood and from the waste. For lumber operations working with logs the operations would be debarking and sawing; for furniture operations they would consist of sawing, planing and sanding. Some of these operations are also point sources if the processing equipment is ducted to a collector. Mechanical cyclone collectors are commonly used. The kilns are also point sources. Table 2.34 shows typical fugitive sources and emissions.

Table 2.34 Typical Woodworking Operations Fugitive Emissions

	Control	
Source	Description	Approximate Efficiency, %
Debarking	Water sprays	50–75
	Enclosure	50
Sawing, planing, sanding	Hood and duct to collector	70–98[a]
Wood waste storage bin	Duct vent to collector	70–98[a]
Wood waste piles	Water sprays	50–75
	Chemical sprays	60–90
	Orient pile	50–70
	Wind screen	60
Roads	Water sprays	50–75
Open areas	Chemical stabilization	80
	Vegetation	65

[a] 70% for cyclone collector and 98% for filter collector depending on dust pickup efficiency.

REFERENCES

1. "The Technical Basis for a Size Specific Particulate Standard, Parts I and II," Specialty Conference, Chatten Cowherd, Jr., Chairman. Edited by the Air Pollution Control Association, APCA, March and April 1980.
2. Wibberley, C. (Compiler), "Proceedings: Fourth Symposium on Fugitive Emissions: Measurement and Control," U.S. EPA No. EPA-600/7-79-182, May 1980.
3. PEDCo Environmental Specialists, "Evaluation of Fugitive Dust Emissions from Mining," U.S. EPA Contract No. 68-02-1321, Task No. 36, April 1976.
4. Brookman, E. T., and D. J. Martin, "A Technical Approach for the Determination of Fugitive Emission Source Strength and Control Requirements," 74th Annual APCA Meeting, Philadelphia, June 1981.
5. Mathai, C. V., L. A. Rathbun and D. C. Drehmel, "An Electrostatically Charged Fog Generator for the Control of Inhalable Particles," EPA Third Symposium on Transfer and Utilization of Particulate Control Technology, Paper No. AV-TP-81/516, Orlando, FL, March 1981.
6. Hoenig, S. A., "University of Arizona Experience in the Control of Dust, Fumes and Smoke by Means of Electrostatically Charged Water Fog," U.S. EPA Pub. No. EPA-600/7-77-131, November 1977.
7. PEDCo Environmental Specialists, "Technical Guidance for Control of Industrial Process Fugitive Particulate Emissions," U.S. EPA No. EPA-450/3-77-010, March 1977.
8. Larson, A. G., et al., "Evaluation of Effectiveness of Civil Engineering Fabrics and Chemical Stabilizers in the Reduction of Fugitive Emissions from Unpaved Roads," 74th Annual APCA Meeting, Paper No. 81-55.4, Philadelphia, June 1981.
9. Chemical Engineering Equipment Buyers' Guide (New York: McGraw Hill Book Company, 1982).
10. Carnes, D., and D. C. Drehmel, "The Control of Fugitive Emissions Using Windscreens," The Third U.S. EPA Symposium on the Transfer and Utilization of Particulate Control Technology, Orlando, FL March 1981.
11. "Industrial Ventilation—A Manual of Recommended Practice," 11th ed. (Lansing, MI Committee on Industrial Ventilation, 1970).
12. Hemeon, W. C. L., "Process Plant Ventilation," 2nd ed., (New York: The Industrial Press, 1963).
13. Dalla Valle, J. M. "Exhaust Hoods," (New York: The Industrial Press, 1946).
14. Silverman, L., "Velocity Characteristics of Narrow Exhaust Slots," *J. of Ind. Hyg. and Toxicol.*, Vol. 24, p. 267, November 1942.
15. Constance, J. D., "Estimating Exhaust-Air Requirements for Processes," *Chem. Eng.*, Vol. 77, No. 17, pp. 116-118, 1970.
16. Castaline, J., "Fugitive Coal Dust Control," *Power Eng.*, pp. 86-87, July 1979.
17. Hesketh, H. E., and F. L. Cross, "The Magnitude of the Agricultural Air Pollution Problem," 1981 Summer Meeting, American Society of Agricultural Engineers, Orlando, FL, June 1981.

18. Evans, J. S., and D. W. Cooper, "An Inventory of Particulate Emissions from Open Winds," *JAPCA*, Vol. 30, No. 12 pp. 1298-1303, December 1980.
19. Richard, G., and D. Sefriet, "Guidelines for Development of Control Strategies in Areas of Fugitive Dust Problems," U.S. EPA No. EPA-450/2-77-029 October 1977.
20. "Compilation of Air Pollutant Emission Factors" AP-42 revised, 1975.
21. Ward, D. E., et al., "An Update on Particulate Emissions from Forest Fires," paper No. 76-2.2, APCA Annual Meeting, Portland, OR, June 1976.
22. McMahon, C. K., and P. W. Ryan, "Some Chemical and Physical Charteristics of Emissions from Forest Fires," paper No. 76-2.3, APCA Annual Meeting, Portland, OR, June 1976.
23. Bohn, R., et al., "Fugitive Emissions from Integrated Iron and Steel Plants," EPA-600/2-78-050, March 1978.
24. PEDCo Environmental Specialists, "Evaluation of Fugitive Dust Emissions from Mining, Task 1 Report, Identification of Fugitive Dust Sources Associated with Mining," U.S. EPA Contract No. 68-02-1321, Task No. 36, April 1976.
25. "Survey of Fugitive Dust from Coal Mines," U.S. EPA No. #PA-908/1-78-003, February 1978.
26. "An Investigation of Particulate Emissions from Construction Aggregate Crushing Operations and Related New Source Performance Standards," NCSA, 1415 Elliot Place, N.W., Washington, D.C., December 1979.
27. Currier, E. L., and B. D. Neal, "Fugitive Emissions from Coal Fired Power Plants," paper No. 70-11.4, proceedings 72nd Annual APCA meeting, Cincinnati, OH, June 1979.
28. "Winds of Change Reshape Air Pollution Modeling," CE, Vol. 87, No. 24, pp. 35-37, December 1, 1980.
29. Hesketh, H. E., and F. L. Cross, "Comparison of Controlled and Uncontrolled Fugitive Emissions from a Large Coal-Fired Utility in Mountainous Terrain," paper No. ICTTE-82-117, The International Congress on Technology and Technology Exchange, Pittsburgh, PA, May 1982.

CHAPTER 3
MEASUREMENT OF EMISSIONS
3.1 FUGITIVE DUST MONITORING

Good emission factors are one of the more important parts of a fugitive emission program. There are two parts to this—obtaining data and standardization of techniques for obtaining data. As more data are gathered, the quality of the emission factors available for estimating fugitive emissions will improve. To obtain even more reliable data, it is necessary to develop a standardization of techniques for monitoring/measuring fugitive emissions.

Fugitive emissions of uncontrolled pollutants emitted into the atmosphere can consist of particulates only or they could also include gases. In general, particulate emissions are more difficult to quantify then fugitive gas emissions. This section discusses particulates only, although some of the techniques are applicable to gases. Particles that are emitted from a point source which discharges emissions through a stack can be readily measured by standard stack testing procedures (e.g., U.S. EPA Method 5 [1]). There are no "standard" techniques for measuring fugitive emissions, but procedures can be used that are recognized as state-of-the-art monitoring techniques for fugitive emissions.

The basic monitoring techniques used are dependent on the type of source and the geometry of the emission source. These methods are:

1. quasi-stack sampling (Section 3.1.2),
2. roof monitor sampling [2] (Section 3.1.3),
3. upwind-downwind sampling [3] (Section 3.1.4),
4. exposure profiling (Section 3.1.5), and
5. tracer sampling (Section 3.1.6).

The first three methods are used more often and therefore are more acceptable than the last two. However, the latter can be useful.

3.1.1 Monitoring Criteria

Before discussing the monitoring techniques, it should be noted that there are three basic types of criteria that must be observed for good monitoring.

Site Criteria

Source Isolability. Can the emissions be measured separately from emissions from other sources? Can the source be enclosed?

Source Location. Is the source indoors or out? Does location permit access of measuring equipment?

Meteorological Conditions. Will wind conditions or precipitation interfere with measurements? Will rain or snow on ground affect dust levels?

Process Criteria

Number and Size of Sources. Are emissions from a single, well defined location or many scattered locations? Is source small enough to hood?

Homogeneity of Emissions. Are emissions the same type everywhere at the site? Are reactive effects between different emissions involved?

Continuity of Process. Will emissions be produced long enough to obtain meaningful samples?

Effects of Measurements. Will installation of measuring equipment alter the process or the emissions? Will measurements interfere with production?

Pollutant Criteria

Nature of Emissions. Are measurements of particles, gases or liquids required? Are emissions hazardous?

Emission Generation Rate. Are enough emissions produced to provide measurable samples in reasonable sampling time?

MEASUREMENT OF EMISSIONS   109

Emission Dilution. Will transport air reduce emission concentration below measurable levels?

3.1.2 Quasi-Stack Sampling

The name implies that this is an almost stack-like sampling technique. A temporary duct or stack is used to collect the sample for true quality-assured stack sampling. This procedure is restricted to emission sources that can be isolated and confined in a manner that provides meaningful data. Care must be taken to prevent loosing the emissions before capture and to prevent excessive dilution by use of too large a collection hood.

It is necessary to withdraw representative and isokinetic samples from the stack according to U.S. EPA Method 1 [1]. Because of the relatively small length-to-diameter ratio that may occur in this type of test setup, it may be necessary to use large numbers of traverse points to compensate for this.

This procedure is ideally suited for single source measurements. Multiple sources could be measured, but only mass emissions would be available unless special physical and/or chemical procedures were used to isolate the particulate catch fractions.

3.1.3 Roof Monitor Sampling

Fugitive emissions may come from a building through roof vents, doors, windows or other openings. It is usually not practicable to attempt to collect all these sources into a single sampling hood nor is it practicable to attempt to sample all openings because of the number and because of the low exit velocities which may make it impossible to use Method 5 techniques and equipment.

The roof monitor sampling technique consists of choosing and sampling at one or more representative roof vents with adequate flow rates so located to provide a representative sample of the particulates. The results obtained cannot be equated with emission factors because they include total building emissions. Also large particulates and heavy gases may not exit through the particular roof monitor chosen. Figure 3.1 shows a possible roof sampling arrangement for a large discharge opening. Some typical measurements from industrial buildings by the roof monitoring technique are listed in Table 3.1. This listing includes both particulates and gases. CO is carbon monoxide, HC is hydrocarbons,

110 FUGITIVE EMISSIONS AND CONTROLS

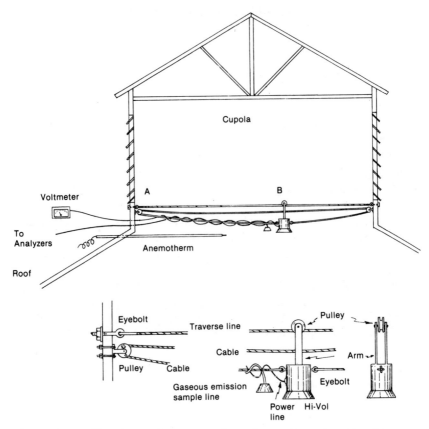

Figure 3.1 Electric arc furnace configuration-roof monitor showing sampling/mounting configuration [2].

SO_2 is sulfur dioxide, HF is hydrogen fluoride (acid) and H_2S is hydrogen sulfide.

3.1.4 Upwind-Downwind Sampling

3.1.4.1 Basic Procedures

This method is applied to large area sources that are not easily confined. This includes roads, storage piles, waste disposal sites and processing operations spread over a large area. The procedure monitors the pollutant levels on both sides of a source with the difference in the two levels representing the contribution from the fugitive source.

Table 3.1 Typical Industrial Fugitive Emissions from Buildings as Measured by the Roof Monitor Sampling Technique [2]

Industry	Source	Particulate Emissions	Gas and Vapor Emissions
Iron and steel foundries	Furnace or cupola charging	Fume, carbon dust, smoke (oil)	CO, HC, SO_2
	Melting	Fume, dust	CO, SO_2
	Mold pouring	Dust	CO, HC, odors
Electric furnace steel	Charging	Metallic fumes, carbon dust	CO
	General operations	Metallic fumes, dust	CO
Primary aluminum	Carbon plant	Tars, carbon dust	CO, HC, SO_2
	Potroom	Tars, carbon and aluminum dust, fluorides	CO, HC, SO_2, HF
	Alumina calcining	Alumina dust	—
	Cryolite recovery	Carbon and alumina dust, fluorides	—
Primary copper	Converter House	Fume, silica	SO_2
	Reverberatory furnace	Fume	SO_2
	Roaster operations	Fume	SO_2
Tires and rubber	Curing press room	Organic particulate	HC, odors
	Cement house	Dust	HC, odors
Phosphate fertilizer	General ventilation	Dust, fluorides	SO_2, HF
Lime	General ventilation	Dust	—
Primary steel	Blast furnace cast house	Metallic fumes	CO, H_2S, SO_2
	BOF operations	Metallic fumes, carbon dust	CO
	Open hearth operations	Metallic fumes	CO
Graphite and carbide production	Arc furnace operation	Carbon dust, silica fume	CO, odors

Upwind-downwind sampling is very dependent on consistent meteorological conditions. The wind direction and speed must be relatively constant during the sampling and the monitors must be oriented appropriately. Temperature, humidity and ground moisture can significantly

112 FUGITIVE EMISSIONS AND CONTROLS

affect the results so the sample must be large enough to minimize these fluctuations.

3.1.4.2 Sector Selector Method

An improved modification of the upwind-downwind technique is the "sector selector method" which is used when there are no predominant sources of emissions upwind of the source involved. Each station upwind and downwind contains two high-volume air samplers (high-vols). One sampler operates a full 24 hr but the second operates only when the wind is blowing from the direction of the source plant. A wind vane can be used to operate the second high-volume air sampler through a sector selector control which turns the unit on when the wind is blowing from the direction of the source through a 22.5° angle. Each unit must have a timer to measure total elapsed time. All monitoring data can be summarized for each of the monitoring periods, and both the background concentration and the contribution from the fugitive source can be assessed. A sketch of a station arrangement is given as Figure 3.2.

3.1.4.3 High-Volume and Other Samplers and Site Locations

Because high-volume samplers are used to obtain the data by this and the prevoius techniques, operation of the samplers, type of samplers and location of the sites are important. The standard U.S. EPA reference method for high-volume samplers is reproduced and included in the Appendix of this book. Actual sampler operation can vary from this in that a continuous 24-hr run may not be desirable, as discussed above. The procedures of accurate weighing, complete drying and careful handling are important and must be followed.

It is preferable to use a type of high-volume sampler with a circular inlet so the orientation direction of the sampler itself is not critical. If old style high-volume samplers are used, the orientation of the inlet with respect to the source and the wind should be noted, because the old gabled roof "dog house" units (TSP type) are direction orientation-specific.

The standard TSP high-volume air sampler collects total suspended particulates (TSP) on an 8- × 10-in. filter (glass fiber or cellulose). This includes particles up to 30 to 100 μ depending on wind speed and direction. Flow rate ranges from 20 to 55 cfm and samples should be from 0.1 to 1 g of particulates. The nongabled-top, nondirectional high-vols can be designed to collect only inhalable particles (up to 15 μ) and are called the IP type. High-volume samplers can be fitted with size-

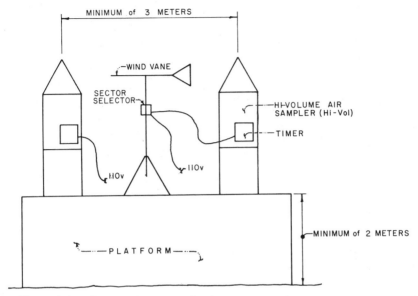

Figure 3.2 Sector selector monitoring station equipment arrangement.

selective cascade impactor heads to provide size and size distribution data as well as mass loadings.

A beta gauge sampler may also be used. This consists of a two-stage collection system of cyclone precollector and beta radiation detector. The cyclone removes the large noninhalable particles, and the smaller particles are counted on a plastic film by carbon-14 beta radiation absorption. The advantage of this system is its ability to measure low dust concentrations (100 to 5000 $\mu g/m^3$).

Site locations are critical with respect to terrain, emission sources and wind velocity. They are also critical with respect to height above the ground. Both ground level and elevated (3 m) stations are recommended at each site if possible. Figure 3.3 shows some suggested site arrangements.

3.1.5 Exposure Profiling

This consists of combining the various measurement techniques to obtain the best profile data for a given source and set of conditions. This is necessary when a number of individual sources exist in one facility and consists of sampling at various locations, times and wind directions.

114 FUGITIVE EMISSIONS AND CONTROLS

Figure 3.3   Typical sampler locations for selected source site locations.

3.1.6 Tracer Sampling

Tracer sampling consists of adding components such as a harmless fluorescent pigment and dye or otherwise "tagged" substance at the source of the emissions. These materials must simulate movement of the dust emissions and must be detectable in very small quantities. In some cases an actual effluent may be used as a tracer. The sampling monitors are examined at the conclusion of the tests to determine how much and from where the sample came, as indicated by the presence of the tracers. This technique is used in combination with the roof monitor and upwind-downwind sampling.

3.1.7 Measurements

After the monitoring techniques have been determined and the sites established, it is necessary to obtain a number of duplicate measurements to statistically achieve meaningful information relative to the desired source while accounting for the numerous parameters mentioned in Section 3.1. Published fugitive emissions data currently available are usually not as reliable as we would like and more data files need to be established for many operations. Zoller, et al. [4] show the inadequacy of industrial emission data. These are shown in Table 3.2. Notice that there is only

Table 3.2   Adequacy of Industrial Emission Data [4]

Industry/Operation	Adequacy of Data[a]
Foundries	VP
Portland cement	P
Minerals extraction/benefication	P
Iron production	P
Secondary lead	VP
Primary aluminum	P
Asphaltic concrete	P
Lime manufacturing	P
Coke manufacturing	F
Secondary aluminum	VP
Secondary brass/bronze	VP
Secondary zinc	VP
Lumber and furniture	P
Concrete batching	F
Primary copper	F
Grain elevators	VG
Primary zinc	VP
Primary lead	F
Steel manufacturing	G

[a]VP = very poor, P = poor, F = fair, G = good, VG = very good.

one very good rating (grain elevators), one good rating (steel manufacturing) and four fair ratings out of the 19 listed.

Mining emission data adequacy is not listed in Table 3.2, but much work has been and is being done to sample and obtain these data. Table 3.3 is an example of a sampling program reported by PEDCo [5] to obtain coal mine fugitive dust emission data. In this program 15 different mining operations were sampled for a total of 213 sampling periods. Even so, fewer than the desired number of samples were obtained for many of the operations, for various reasons, during this program.

Table 3.3 Coal Mine Operations Fugitive Dust Samples [5]

	Number of Sampling Periods at Mine[a]					
Operation	A	B	C	D	E	Total
Dragline	8	10	6	6	N/A	30
Haul road traffic	10	10	10	9	8	47
Shovel/truck loading	6	6	4	0	10	26
Blasting: overburden	1	N/A	2	N/A	2	13
coal	N/A	2	2	2	2	
Truck dump	6	2	4	6	4	22
Storage piles	6	8	4	4	N/A	22
Exposed areas	4	6	4	3	4	21
Drilling	1	0	2	0	2	5
Fly ash dump	3	0	N/A	N/A	N/A	3
Train loading	N/A	N/A	4	N/A	5	9
Topsoil removal	0	0	0	10	0	10
Front end loaders	N/A	N/A	N/A	1	N/A	1
Graders	0	N/A	N/A	0	N/A	0
Bucket wheel	N/A	N/A	N/A	N/A	0	0
Overburden dump	N/A	N/A	N/A	N/A	4	4
Total	45	44	42	41	41	213

[a]N/A = not applicable.

3.2 ESTIMATING

3.2.1 Estimating Fugitive Emissions

Emission factors have been established for many fugitive sources of emissions, as presented in Section 1.6. The emission factor when multiplied by the production rate or throughput of material will provide an estimate of the atmospheric emission. A typical example of this type of datum is the summary of emission estimates for various types of mining operations listed in Table 3.4 [6]. In this table there is no information

Table 3.4 Example Summary of Emission Estimates for Mining Operations [6]

Operation	No. of Emission Estimates	Range	Units	Coal	Copper	Rock	P_2O_5 Rock[a]	More Data Needed
Overburden removal	5	0.0008–0.45 0.048–0.10	lb/ton of ore lb/ton of overburden	0.05	0.0008			x
Blasting	2	0.001–0.16	lb/ton of ore					x
Shovels/truck loading	5	negligible–0.10	lb/ton of ore	0.05	0.05	0.05	N/A	
Haul roads	4	0.8–2.2	lb/vehicle mile traveled		depends on speeds and controls			x
Truck dumping	3	0.00034–0.04	lb/ton of ore	0.02	0.02	0.04	N/A	
Crushing	4	negligible–0.7	lb/ton of ore	neg		0.044		x
Transfer and conveyance	5	negligible–0.2	lb/ton of ore				0.15	x
Cleaning	0	usually negligible			neg		neg	
Storage	5	0.0235–0.42 3.5–13.2	lb/ton of ore lb/acre/day	0.054	0.33	0.33 10.4	0.20	
Waste disposal	1	negligible–14.4	ton/acre/yr					
Reclamation	1	use wind erosion equation	ton/acre/yr		depends on climate and soil			x

[a] N/A = not applicable.

118 FUGITIVE EMISSIONS AND CONTROLS

on the reliability of these estimates. Instead the table lists how many estimates were used in arriving at this number.

The steps involved in estimating fugitive emissions at a plant consist of:

1. location of the sources of emissions from a plant plot plan and flow diagram.
2. determination of the throughput of material at each point (lb/hr, tons/yr).
3. selection of appropriate emission factors, and
4. the calculation and tabulation of emissions for each source of emission.

The following example for a cement plant will provide an insight, on an actual example, as to how fugitive emissions are estimated. Included are location of the source, selections of emission factors and calculations of the particulate emissions.

Subject example is for an 800,000-ton per year (TPY) cement plant. Applying BACT, the emission summary indicated a total emission of 271 TPY as follows:

$$\begin{aligned}\text{Point source emissions} &= 269.88 \text{ TPY} \\ \text{Fugitive sources} &= \underline{1.41 \text{ TPY}} \\ \text{Total emissions} &= 271.29 \text{ TPY}\end{aligned}$$

Fugitive sources are located in the following areas (see Figure 3.4):

- limestone handling,
- iron ore handling,
- fly ash handling,
- coal handling,

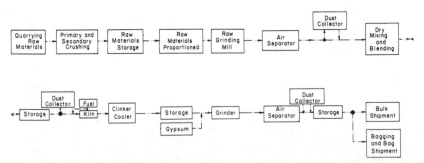

Figure 3.4 Basic flow diagram of Portland cement manufacturing process.

- kiln bypass dust handling, and
- gypsum handling.

The controlled emissions summarized in Table 3.5 are the result of applying Midwest Research Institute (MRI) fugitive emission formulas.

Table 3.5   Example 800,000-TPY Cement Plant Emission Summary for Fugitive Sources

Source	Operating Schedule			Emissions	
	hr/day	days/wk	wk/yr	lb/hr	TPY
LIMESTONE HANDLING SYSTEM					
T-1 Quarry to A01	8	5	47	28.00	26.30
T-2 Conveyor A01 to Hopper B01	8	5	47	zero	zero
T-3 Hopper B01 to Crusher	8	5	47	zero	zero
T-4 Crusher to C01	8	5	47	zero	zero
T-5 C01 to Limestone Pile	8	5	47	zero	zero
T-6 Limestone Pile to D01	22	7	47	zero	zero
T-7 D01 to Limestone Feed Bin D02	22	7	47	zero	zero
T-8 Limestone Feed Bin to Conveyor E03	22	7	47	zero	zero
IRON ORE HANDLING Pile (T-9) Fe Storage Pile	4	7	48	2.5	1.68
T-10 Front End Loader to Elevator D05	4	7	48	0.74	0.50
T-11 Elevator D05 to Fe Feed Bin	4	7	48	zero	zero
T-12 Fe Feed Bin to Conveyor E03	4	7	48	zero	zero

Table 3.5, continued

Source	Operating Schedule			Emissions	
	hr/day	days/wk	wk/yr	lb/hr	TPY
PEA GRAVEL HANDLING					
T-13 Gravel Storage Pile	4	7	48	2.50	1.68
T-14 Loader to Elevator D05	4	7	48	0.74	0.50
T-15 Elevator D05 to Gravel Feed Bin	4	7	48	zero	zero
T-16 Gravel Feed Bin to Conveyor E03	4	7	48	zero	zero
FLY ASH HANDLING					
T-17 Truck to Bin D09	22	7	48	zero	zero
T-18 Bin to Conveyor E03	22	7	48	zero	zero
RAW MATERIAL HANDLING					
T-19 Conveyor E03 to Raw Mill	22	7	48	zero	zero
COAL HANDLING					
T-20 Rail car to Conveyor C06	4	7	48	10.00	6.70
T-21 Coal Storage Pile	4	7	48	6.30	4.20
T-22 Loader to Conveyor S04	4	7	48	1.60	1.10
T-23 Conveyor S04 Coal Mill Feed Bin	4	7	48	4.00	2.70
T-24 Bin to Weigh Feeder	24	7	48	0.60	2.40

MEASUREMENT OF EMISSIONS

Table 3.5, continued

Source	Operating Schedule			Emissions	
	hr/day	days/wk	wk/yr	lb/hr	TPY
KILN BYPASS DUST HANDLING					
T-25 Weigh Feeder to Grinding Building	24	7	48	zero	zero
T-26 Conveyor K27 to Elevator K28	24	7	48	zero	zero
T-27 Elevator K28 to Conveyor K29	24	7	48	zero	zero
T-28 Conveyor K29 to Bin K30	24	7	48	zero	zero
T-29 Bin to Truck	24	7	48	zero	zero
T-30 Pug Mill to Truck	4	7	48	negligible	negligible
GYPSUM HANDLING					
T-31 Railcar to Conveyor C06	4	7	48	10.00	6.70
T-32 Gyp pile Emissions	4	7	48	6.30	4.20
T-33 Loader to Conveyor L15	2	7	48	3.90	1.30
T-34 L15 to Bin L16	2	7	48	zero	zero
				TOTAL	59.96

A typical calculation example would be for the first transfer point in the limestone handling system, as follows:

T-1 Process rate = 1,316,000 TPY

$$\text{Emission factor (MRI)} = 0.0018 \frac{\left(\frac{S}{5}\right)\left(\frac{u}{5}\right)}{\left(\frac{M}{2}\right)^2 \left(\frac{Y}{6}\right)} \text{ lb/ton}$$

$S = 1.5\%$ $M = 7.5\%$
$u = 6.7$ mph $Y = 2$ yd^3

$$EF = 0.0018 \frac{\left(\frac{1.5}{5}\right)\left(\frac{6.7}{5}\right)}{\left(\frac{7.5}{2}\right)^2\left(\frac{2}{6}\right)} = 155.9 \times 10^{-6}$$

$$\text{Emissions} = \frac{(155.9 \times 10^{-6})(1.316 \times 10^6)}{2000} \text{ TPY}$$

$$E = 0.102 \text{ TPY}$$

The summary of these emissions based on application BACT to the uncontrolled fugitive emissions is presented in Table 3.6.

3.2.2 Estimating Control Efficiencies

Control efficiency estimates are needed to be able to specify emissions from various sources and combinations of sources and to develop practi-

Table 3.6 Summary of Fugitive Emissions with BACT Controls from an 800,000-TPY Cement Plant

Source	Uncontrolled Emission, (lb/hr)[a]	Type of Control Device	Degree of Control, %	Emission, lb/hr
LIMESTONE HANDLING SYSTEM				
T-1	140.0	Wet Material	80	28.00
T-2	NE	Enclosed	100	zero
T-3	NE	Enclosed	100	zero
T-4	NE	Enclosed	100	zero
T-5	NE	Enclosed	100	zero
T-6	NE	Enclosed	100	zero
T-7	NE	Enclosed	100	zero
T-8	NE	Enclosed	100	zero
IRON ORE HANDLING				
T-9	25.0	Sprays	90	2.50
T-10	3.7	Enclosed	80	0.74
T-11	NE	Enclosed	100	zero
T-12	NE	Enclosed	100	zero
T-13	25.0	Sprays	90	2.50
T-14	3.7	Partially Enclosed	80	0.74
T-15	NE	Enclosed	100	zero
T-16	NE	Enclosed	100	zero

Table 3.6, continued

Source	Uncontrolled Emission, (lb/hr)[a]	Type of Control Device	Degree of Control, %	Emission, lb/hr
FLY ASH HANDLING				
T-17	NE	Enclosed	100	zero
T-18	NE	Enclosed	100	zero
RAW MATERIAL HANDLING				
T-19	NE	Enclosed	100	zero
COAL HANDLING				
T-20	100.0	Enclosed	90	10.00
T-21	63.0	Sprays	90	6.30
T-22	8.0	Partially Enclosed	80	1.60
T-23	8.0	Wet Material	50	4.00
T-24	1.2	Wet Material	50	0.60
T-25	NE	Enclosed	100	zero
KILN BYPASS DUST HANDLING				
T-26	NE	Enclosed	100	zero
T-27	NE	Enclosed	100	zero
T-28	NE	Enclosed	100	zero
T-29	NE	Enclosed	100	zero
T-30	NE	Enclosed	100	zero
GYPSUM HANDLING				
T-31	100.0	Enclosed	90	10.00
T-32	63.0	Enclosed	90	6.30
T-33	7.8	Partially Enclosed	50	3.90
T-34	NE	Enclosed	100	zero

[a]NE = not estimated.

cable and economic control strategies needed to meet the air quality requirements. It may be possible to utilize emission factor data such as in Section 1.6, or it may be necessary to obtain actual measurements of existing emissions. In order to obtain estimates for controlled emissions after application of various controls and control combinations, it may be necessary to conduct model studies either in the field or in a laboratory wind tunnel if such emission data are not available. Once these data are available, control efficiencies can be estimated. Several examples are presented below. Note the precautions needed in these procedures.

124 FUGITIVE EMISSIONS AND CONTROLS

Unpaved Road, Example 1. Estimate control efficiency of a 1-in. chip seal paving using field measured data of 304 $\mu g/m^3$ from a control section of road, 88 $\mu g/m^3$ from a test paved section and 50 $\mu g/m^3$ as background ambient dust level

$$\% \text{ control} = \frac{(304 - 50) - (88 - 50)}{(304 - 50)} \, 100 = 85\%$$

Unpaved Road, Example 2. Estimate control efficiency of a watered road. Data show total emissions of 8.5 lb/vehicle mile traveled on a dry road at 50 mph. Over the same period wind erosion produced the equivalent of 2.1 lb, although no vehicle traveled on the road. Watered road data gave emission values of 4.7 for 25 mph and 0.5 for no vehicle.

$$\% \text{ control} = \frac{(8.5 - 2.1) - (4.7 - 0.5)}{(8.5 - 2.1)} \, 100 = 34\%$$

Windbreak Example. Estimate control efficiency of a windbreak. Data show this is not a single number. Depending on windbreak height (and other features) and measurements locations, wind speed reductions range as shown by Figure 2.3 from 30 to 110% of incoming wind speed. Where emissions are a direct function of wind speed, control efficiencies would range from 70 to a negative 10%.

3.2.3 Models

There are numerous equations and modeling procedures developed to utilize fugitive emission data and correlate these with appropriate meteorological and geographic data, once they are available, to estimate downwind ambient concentrations. For example, emissions may be from an area source, such as an industrial area with many different sources, or they may be considered as coming from a long line of sources when the wind direction is normal to that line. For ground level emissions, the dispersion equations would be for a line source:

$$C_{(x,0,0)} = \frac{2q_1}{\sin \phi \sqrt{2\pi} \, \sigma_z u}$$

where $C_{(x,0,0)}$ = plume centerline concentration at ground level a distance x downwind from the source, g/m³
  q_1 = line source strength g/(sec m)
  σ_z = vertical dispersion coefficient, m (e.g., see Turner [7])
  ϕ = angle between wind direction and line source, degrees
  u = mean wind speed, m/sec

For an area source, the dispersion equation is:

$$C_{(x,0,0)} = \frac{q_2}{\pi \, \sigma_y \, \sigma_z \, u}$$

where σ_y = horizontal dispersion coefficient, m (e.g., see Turner [7])
  q_2 = area source strength, g/sec

In addition, settling of the particles must be accounted for using some tilted plume or fallout function adaptation.

There are computer models developed by the U.S. EPA, universities and others to account for these variations but each must be used with care. The U.S. EPA VALLEY model was used in the Section 2.3.9 example. This is similar to Martin's model [8], with the inclusion of factors to account for mountainous downwind terrain. The VALLEY model utilizes Briggs' plume rise functions [9] and estimates emissions in fixed, circular network receptors along radials from the source. Some plume height above ground is required (usually 3 m). Multiple sources require repetition of the program with superimposition of concentrations on the computer printouts. As noted, currently the VALLEY model is the only model accepted by the U.S. EPA for high terrain studies.

Other significant diffusion models are the CRSTER and RAM models. These models are recommended by the U.S. EPA [10] and widely used. However, they predict substantially different downwind pollutant concentrations. Martin [11] has recommended changes to these models.

All procedures are subject to local factors which influence dust deposition as noted by El-Shobokshy and Eldin [12]. Their findings show that for stable atmosphere, particles 0.1 to 100 μ fall out of the air at about their gravitational settling velocity. Particles 0.1 to 10 μ are highly affected by neutral and unstable atmospheres. For example, in an unstable atmosphere in an urban area (Z $\simeq$ 5m) at 4 m/sec wind speeds, a 10-μ particle deposits 40 times more rapidly than in the open country under the same conditions. This is shown in Figure 3.5.

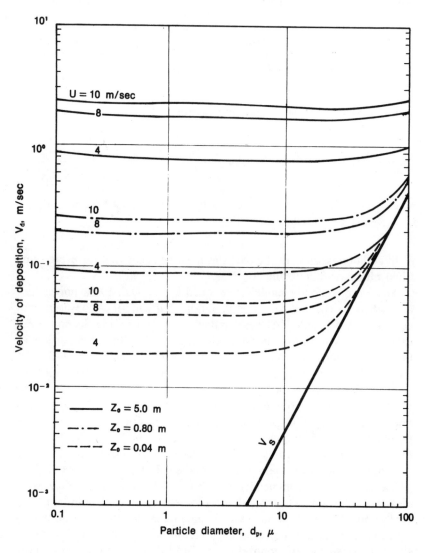

Figure 3.5 Velocity of deposition of dust particles at various wind speeds and surface roughness for unstable atmosphere [12].

where Z_o is surface roughness in m (≤ 0.06 m for open country; ≤ 1.0 m for forested areas); ≤ 10 m for urban areas),
V_s is terminal settling velocity in m/sec,
V_d is particle deposition velocity in m/sec, and
u is wind speed in m/sec

Table 3.7 Quarterly Suspended Particulate Report

Company Submitting Report: _____

Monitoring Period: _____

Station Location: _____

Summary of Field Data:

Date	24-Hour Reading	Contribution by This Company, $\mu g/m^3$	Background $\mu g/m^3$	Wind Direction, Degrees	Wind Speed, mph	Comments

*Average value for period: _____ $\mu g/m^3$
**Maximum value recorded: _____ $\mu g/m^3$
***Second highest reading: _____ $\mu g/m^3$

3.3 REPORTING FUGITIVE EMISSIONS

It may be desirable to monitor fugitive emissions by use of high-volume samplers located on or around the perimeter of a facility. It may also be necessary to report these findings. Table 3.7 is a proposed quarterly particulate report and Table 3.8 is a proposed periodic violations report. It may be necessary to modify these forms if size-specific data are required—for example, % $\leq$15 μ and % $\leq$2.5 μ for inhalable and respirable fractions respectively.

Table 3.8 Period Air Quality Violations Report
(violation of primary air quality standard at 250 μg/m³)

Date	Station Location	Station Value, μg/m³	Contribution by This Company, μg/m³	Comments

Company Submitting Report: _____
Data Collected By _____
Laboratory Analysis By: _____
Report Prepared By: _____
(Signature)

REFERENCES

1. "Standards of Performance for New Stationary Sources," U.S. EPA publication No. EPA-340/1-77-015, Title 40, 60, as amended August 18, 1977.
2. "Technical Manual for the Measurement of Fugitive Emissions: Roof Monitor Sampling Method for Industrial Fugitive Emissions," U.S. EPA publication No. EPA-600 2-76-089b, May 1976.
3. "Technical Manual for Measurement of Fugitive Emissions: Upwind/Downwind Sampling Method for Industrial Emissions," U.S. EPA publication No. EPA-600/2-76-089a, April 1976.

4. Zoller, J., T. Bertice and T. Janszen, "Assessment of Fugitive Particulate Emission Factors for Industrial Processes," U.S. EPA publication No. EPA-450/3-78-107, September 1978.
5. "Survey of Fugitive Dust from Coal Mines," PEDCo Environmental Specialists, U.S. EPA Contract No. 68-01-4489, Project No. 3311, February 1978.
6. "Evaluation of Fugitive Dust Emissions from Mining, Task 1 Report, Identification of Fugitive Dust Sources Associated with Mining," PEDCo Environmental Specialists, Inc., U.S. EPA Contract No. 68-02-1321, Task No. 36, April 1976.
7. Turner, D. B., "Workbook of Atmospheric Dispersion Estimates," U.S. Dept. HEW, Revised 1969.
8. Martin, D. O., "An Urban Diffusion Model for Estimating Long Term Average Values for Air Quality," *JAPCA* Vol. 21, No. 1, pp. 16-19, January 1971.
9. Briggs, G. A., "Plume Rise," AEC Critical Review Series, TID-25075, 1969.
10. "Guidelines on Air Quality Models," U.S. EPA publication No. EPA-450/2-78-027, April 1978.
11. Martin, J. R., "Critical Inconsistencies Between the Official U.S. EPA Rural Version of the CRSTER and RAM Diffusion Models," paper No. 79-12.3, 72nd Annual APCA Meeting, Cincinnati, OH, June 1979.
12. El-Shobokshy, M. S., and A. S. Eldin, "Estimation of Dust Deposits from the Atmosphere over Open Areas at Different Atmospheric Stabilities," University of Riyadh, private communication, October 1981.

APPENDIX

REFERENCE METHOD FOR THE DETERMINATION OF SUSPENDED PARTICULATES IN THE ATMOSPHERE (HIGH-VOLUME METHOD)*

1. Principle and Applicability

1.1 Air is drawn into a covered housing and through a filter by means of a high-flow-rate blower at a flow rate (1.13 to 1.70 m^3/min; 40 to 60 ft^3/min) that allows suspended particles having diameters of less than 100 μ (Stokes equivalent diameter) to pass to the filter surface. [1] Particles within the size range of 100 to 0.1 μ diameter are ordinarily collected on glass fiber filters. The mass concentration of suspended particulates in the ambient air (μg/m^3) is computed by measuring the mass of collected particulates and the volume of air sampled.

1.2 This method is applicable to measurement of the mass concentration of suspended particulates in ambient air. The size of the sample collected is usually adequate for other analyses.

2. Range and Sensitivity

2.1 When the sampler is operated at an average flow rate of 1.70 m^3/min (60 ft^3/min) for 24 hr, an adequate sample will be obtained even in an atmosphere having concentrations of suspended particulates as low as 1 μg/m^3. If particulate levels are unusually high, a satisfactory sample may be obtained in 6 to 8 hr or less. For determination of average con-

---

*Reproduced from PA. LA. 56.2.73, Federal Register, Vol. 36, No. 84; Friday, April 30, 1971, and originally published as Code of Federal Regulation 40, Part 50.11, Appendix B, July 1, 1975, pp. 12–16.

centrations of suspended particulates in ambient air, a standard sampling period of 24 hr is recommended.

2.2 Weights are determined to the nearest mg, air flow rates are determined to the nearest 0.03 m^3/min (1.0 ft^3/min); times are determined to the nearest 2 min, and mass concentrations are reported to the nearest μg/m^3.

3. Interferences

3.1 Particulate matter that is oily, such as photochemical smog or wood smoke, may block the filter and cause a rapid drop in air flow at a nonuniform rate. Dense fog or high humidity can cause the filter to become too wet and severely reduce the air flow through the filter.

3.2 Glass-fiber filters are comparatively insensitive to changes in relative humidity, but collected particulates can be hygroscopic [2].

4. Precision, Accuracy and Stability

4.1 Based on collaborative testing, the relative standard deviation (coefficient of variation) for single analyst variation (repeatability of the method) is 3.0%. The corresponding value for multilaboratory variation (reproducibility of the method) is 3.7% [3].

4.2 The accuracy with which the sampler measures the true average concentration depends on the constancy of the air flow rate through the sampler. The air flow rate is affected by the concentration and the nature of the dust in the atmosphere. Under these conditions the error in the measured average concentration may be in excess of ±50% of the true average concentration, depending on the amount of reduction of air flow rate and on the variation of the mass concentration of dust with time during the 24-hr sampling period [4].

5. Apparatus

5.1 Sampling

5.1.1 Sampler. The sampler consists of three units: (1) the face plate and gasket, (2) the filter adapter assembly, and (3) the motor unit. The sampler must be capable of passing environmental air through a 406.5 cm^2 (63 in.^{2}) portion of a clean 20.3- × 25.4-cm (8- × 10-in.) glass-fiber filter at a rate of at least 1.70 m³/min. (60 ft³/min). The motor must be capable of continuous operation for 24-hr periods with input voltages ranging from 110 to 120 V, 50–60 Hz alternating current, and must have a third-wire safety ground. The housing for the motor unit may be of any convenient construction so long as the unit remains airtight and leak-free. The life of the sampler motor can be extended by lowering the voltage by about 10% with a small "buck or boost" transformer between the sampler and power outlet.

5.1.2 Sampler Shelter. It is important that the sampler be properly installed in a suitable shelter. The shelter is subjected to extremes of temperature, humidity, and all types of air pollutants. For these reasons the materials of the shelter must be chosen carefully. Properly painted exterior plywood or heavy-gauge aluminum serve well. The sampler must be mounted vertically in the shelter so that the glass-fiber filter is parallel with the ground. The shelter must be provided with a roof so that the filter is protected from precipitation and debris. The clearance area between the main housing and the roof at its closest point should be 580.5 ± 193.5 cm² (90 ± 30 in.²). The main housing should be rectangular, with dimensions of about 29 × 36 cm (11.5 × 14 in.).

5.1.3 Rotameter. Marked in arbitrary units, frequently 0 to 70, and capable of being calibrated. Other devices of at least comparable accuracy may be used.

5.1.4 Orifice Calibration Unit. Consisting of a metal tube 7.6 cm (3 in.) i.d. and 15.9 cm (6¼ in.) long with a static pressure tap 5.1 cm (2 in.) from one end. The tube end nearest the pressure tap is flanged to about 10.8 cm (4¼ in.) o.d. with a male thread of the same size as the inlet end of the high-volume air sampler. A single metal plate 9.2 cm (3⅝ in.) in diameter and 0.24 cm (3/32 in.) thick with a central orifice 2.9 cm (1⅛ in.) in diameter is held in place at the air inlet end with a female threaded ring. The other end of the tube is flanged to hold a loose female threaded coupling, which screws onto the inlet of the sampler. An 18-hole metal plate, an integral part of the unit, is positioned between the orifice and sampler to simulate the resistance of a clean glass-fiber filter.

5.1.5 Differential Manometer. Capable of measuring to at least 40 cm (16 in.) of water.

5.1.6 Positive Displacement Meter. Calibrated in cubic meters or cubic feet, to be used as a primary standard.

5.1.7 Barometer. Capable of measuring atmospheric pressure to the nearest mm.

5.2 Analysis

5.2.1 Filter Conditioning Environment. Balance room or desiccator maintained at 15 to 35°C and less than 50% relative humidity.

5.2.2 Analytical Balance. Equipped with a weighing chamber designed to handle unfolded 20.3 × 25.4-cm (8- × 10-in.) filters and having a sensitivity of 0.1 mg.

5.2.3 Light Source. Frequently a table of the type used to view X-ray films.

5.2.4 Numbering Device. Capable of printing identification numbers on the filters.

6. Reagents

6.1 Filter Media. Glass-fiber filters having a collection efficiency of at least 99% for particles of 0.3 μ diameter, as measured by the dioctyl-phthalate (DOP) test, are suitable for the quantitative measurement of concentrations of suspended particulates [5], although some other medium, such as paper, may be desirable for some analyses. If a more detailed analysis is contemplated, care must be exercised to use filters that contain low background concentrations of the pollutant being investigated. Careful quality control is required to determine background values of these pollutants.

7. Procedure

7.1 Sampling

7.1.1 Filter Preparation. Expose each filter to the light source and inspect for pinholes, particles or other imperfections. Filters with visible imperfections should not be used. A small brush is useful for removing particles. Equilibrate the filters in the filter conditioning environment for 24 hr. Weigh the filters to the nearest mg; record tare weight and filter identification number. Do not bend or fold the filter before collection of the sample.

7.1.2 Sample Collection. Open the shelter, loosen the wing nuts, and remove the face plate from the filter holder. Install a numbered, pre-weighed, glass fiber filter in position (rough side up), replace the face plate without disturbing the filter, and fasten securely. Undertightening will allow air leakage, overtightening will damage the sponge-rubber face-plate gasket. A very light application of talcum powder may be used on the spronge-rubber face-plate gasket to prevent the filter from sticking. During inclement weather the sampler may be removed to a protected area for filter change. Close the roof of the shelter, run the sampler for about 5 min, connect the rotameter to the nipple on the back of the sampler, and read the rotameter ball with the rotameter in a vertical position. Estimate to the nearest whole number. If the ball is fluctuating rapidly, tip the rotameter and slowly straighten it until the ball gives a constant reading. Disconnect the rotameter from the nipple; record the initial rotameter reading and the starting time and date on the filter folder. (The rotameter should never be connected to the sampler except when the flow is being measured.) Sample for 24 hr from midnight to midnight and take a final rotameter reading. Record the final rotameter reading and ending time and date on the filter folder. Remove the face-plate as described above and carefully remove the filter from the holder, touching only the outer edges. Fold the filter lengthwise so that only surfaces with collected particulates are in contact, and place in a manila folder. Record on the folder the filter number, location, and any other factors, such as meteorological conditions or razing of nearby buildings, that might affect the results. If the sample is defective, void it at this time. In order to obtain a valid sample, the high-volume sampler must be operated with the same rotameter and tubing that were used during its calibration.

7.2 Analysis

Equilibrate the exposed filters for 24 hr in the filter conditioning environment, then reweigh. After they are weighed, the filters may be saved for detailed chemical analysis.

7.3 Maintenance

7.3.1 Sampler Motor. Replace brushes before they are worn to the point where motor damage can occur.

7.3.2 Face-Plate Gasket. Replace when the margins of samples are no longer sharp. The gasket may be sealed to the face-plate with rubber cement or double-sided adhesive tape.

136 FUGITIVE EMISSIONS AND CONTROLS

7.3.3 Rotameter. Clean as required, using alcohol.

8. Calibration

8.1 Purpose

Since only a small portion of the total air sampled passes through the rotameter during measurement, the rotameter must be calibrated against actual air flow with the orifice calibration unit. Before the orifice calibration unit can be used to calibrate the rotameter, the orifice calibration unit itself must be calibrated against the positive displacement primary standard.

8.1.1 Orfice Calibration Unit. Attach the orifice calibration unit to the intake end of the positive displacement primary standard and attach a high-volume motor blower unit to the exhaust end of the primary standard. Connect one end of a differential manometer to the differential pressure tap of the orifice calibration unit and leave the other end open to the atmosphere. Operate the high-volume motor blower unit so that a series of different, but constant, air flows (usually six) are obtained for definite time periods. Record the reading on the differential manometer at each air flow. The different constant air flows are obtained by placing a series of load plates, one at a time, between the calibration unit and the primary standard. Placing the orifice before the inlet reduces the pressure at the inlet of the primary standard below atmospheric; therefore, a correction must be made for the increase in volume caused by this decreased inlet pressure. Attach one end of a second differential manometer to an inlet pressure tap of the primary standard and leave the other open to the atmosphere. During each of the constant air flow measurements made above, measure the true inlet pressure of the primary standard with this second differential manometer. Measure atmospheric pressure and temperature. Correct the measured air volume to true air volume as directed in 9.1.1, then obtain true air flow rate, Q, as directed in 9.1.3. Plot the differential nanometer readings of the orifice unit vs. Q.

8.1.2 High-Volume Sampler. Assemble a high-volume sampler with a clean filter in place and run for at least 5 min. Attach a rotameter, read the ball, adjust so that the ball reads 65, and seal the adjusting mechanism so that it cannot be changed easily. Shut off motor, remove the filter, and attach the orifice calibration unit in its place. Operate the high-volume sampler at a series of different, but constant, air flows (usually six). Record the readings of the differential manometer on the orifice calibration unit, and record the readings of the rotameter at each flow. Measure atmospheric pressure and temperature. Convert the differential manometer reading to m³/min, Q, then plot rotameter reading vs Q.

8.1.3 Correction for Differences in Pressure or Temperature. See Addendum B.

9. Calculations

9.1 Calibration of Orifice

9.1.1 True Air Volume. Calculate the air volume measured by the positive displacement primary standard.

$$V_a = \frac{(P_a - P_m)}{P_a}(V_M)$$

where V_a = true air volume at atmospheric pressure, m³
 P_a = barometric pressure, mm Hg
 P_m = pressure drop at inlet of primary standard, mm Hg
 V_M = volume measured by primary standard, m³

9.1.2. Conversion Factors.

$$\text{in. Hg} \times 25.4 = \text{mm Hg}$$
$$\text{in. water} \times 73.48 \times 10^{-3} = \text{in. Hg}$$
$$\text{ft}^3 \text{ air} \times 0.0284 = \text{m}^3 \text{ air}$$

9.1.3 True Air Flow Rate.

$$Q = \frac{V_a}{T}$$
Q = flow rate, m³/min
T = time of flow, min

9.2 Sample Volume

9.2.1 Volume Conversion. Convert the initial and final rotameter readings to true air flow rate, Q, using the calibration curve of 8.1.2.

9.2.2 Calculate volume of air sampled

$$V = \frac{Q_i + Q_f}{2} \times T$$

where V = air volume sampled, m³
Q_i = initial air flow rate, m³/min
Q_f = final air flow rate, m³/min
T = sampling time, min

9.3 Calculate mass concentration of suspended particulates

$$S.P. = \frac{(W_f - W_i) \times 10^6}{V}$$

where S.P. = mass concentration of suspended particulates, $\mu g/m^3$
W_i = initial weight of filter, g
W_f = final weight of filter, g
V = air volume sampled, m³
10^6 = conversion of g to μg

10. References

1. Robson, C. D., and K. E. Foster, "Evaluation of Air Particulate Sampling Equipment," *Am. Ind. Hyg. Assoc. J.* 24, 404 (1962).
2. Tierney, G. P., and W. D. Conner, "Hygroscopic Effects on Weight Determinations of Particulates Collected on Glass-Fiber Filters," *Am. Ind. Hyg. Assoc. J.* 28, 363 (1967).
3. Unpublished data based on a collaborative test involving 12 participants, conducted under the direction of the Methods Standardization Services Section of the National Air Pollution Control Administration, October, 1970.
4. Harrison, W. K., J. S. Nader, and S. F. Fugman, "Constant Flow Regulators for High-Volume Air Sampler," *Am. Ind. Hyg. Assoc. J.* 21, 114-120 (1960)
5. Pate, J. B., and E. C. Tabor, "Analytical Aspects of the Use of Glass-Fiber Filters for the Collection and Analysis of Atmospheric Particulate Matter," *Am. Ind. Hyg. Assoc. J.* 23, 144-150 (1962).

ADDENDA

A. Alternative Equipment

A modification of the high-volume sampler incorporating a method for recording the actual air flow over the entire sampling period has been described, and is acceptable for measuring the concentration of suspended particulates (Henderson, J. S., Eighth Conference on Methods in Air Pollution and Industrial Hygiene Studies, 1967, Oakland, CA). This modification consists of an exhaust orifice meter assembly connected through a transducer to a system for continuously recording air flow on a circular chart. The volume of air sampled is calculated by the following equation:

$$V = Q \times T$$

where Q = average sampling rate, m^3/min
T = sampling time, min

The average sampling rate, Q, is determined from the recorder chart by estimation if the flow rate does not vary more than 0.11 m^3/min. (4 ft^3/min) during the sampling period. If the flow rate does vary more than 0.11 m^3 (4 ft^3/min) during the sampling period, read the flow rate from the chart at 2-hr intervals and take the average.

B. Pressure and Temperature Corrections

If the pressure or temperature during high-volume sampler calibration is substantially different from the pressure or temperature during orifice calibration, a correction of the flow rate, Q, may be required. If the pressures differ by no more than 15% and the temperatures differ by no more than 100% (°C), the error in the uncorrected flow rate will be no more than 15%. If necessary, obtain the corrected flow rate as directed below. This correction applies only to orifice meters having a constant orifice coefficient. The coefficient for the calibrating orifice described in 5.1.4 has been shown experimentally to be constant over the normal operating range of the high-volume sampler (0.6 to 2.2 m^3/min; 20 to 78 ft^3/min). Calculate corrected flow rate:

$$Q_2 = Q_1 \left(\frac{T_2 P_1}{T_1 P_2}\right)^{1/2}$$

where Q_2 = corrected flow rate, m³/min
Q_1 = flow rate during high-volume sampler calibration (Section 8.1.2), m³/min
T_1 = absolute temperature during orifice unit calibration (Section 8.1.1), °K or °R
P_1 = barometric pressure during orifice unit calibration (Section 8.1.1), mm Hg
T_2 = absolute temperature during high-volume sampler calibration (Section 8.1.2), °K or °R
P_2 = barometric pressure during high-volume sampler calibration (Section 8.1.2), mm Hg

NOMENCLATURE

A = area
Å = angstrom unit, 10^{-10}m
C_D = drag coefficient, dimensionless
$C_{X,o,o}$ = centerline, ground level downwind concentration, g/m^3
d = particle diameter
d_A = adjusted diameter
D_1 = duration of material storage, days
D_2 = dry days per year
e = surface erodibility
E = emission factor
E_{30} = emission factor for dust <30 μ
f = percent of time wind exceeds 12 mph
g = gravitational acceleration or grams
ha = hectare, 2.471 acres, 10,000 m^2
kg = kilogram
K = vehicular activity factor
  = 0 for no traffic
  = 0.25 for iron ore pellets, coal and large stone
  = 1.0 for screened stone
K_v = von Karmen constant
L_1 = surface dust loading
m = meter
M = material moisture content, % by mass
pCi = pico Curie, 10^{-12} Curie
PE = Thornthwaite's precipitation-evaporation index
Q = volumetric flow rate, cfm
q = source strength
Re = Reynolds number, dimensionless
s = silt content, %
S = vehicle speed
T = number of tillings per year
U = mean wind speed
U_f = terminal velocity
U_{*_t} = threshold friction velocity
v = air velocity
V = volume
VMT = vehicle mile traveled
w = mean annual days of $\geq$0.01 in. rainfall
W = vehicle weight

x = distance
Y = loader bucket capacity
z = height
z_o = roughness parameter

GREEK SYMBOLS

α = dimensionless sheer-to-stress ratio
ρ_a = air density
ρ_b = bulk density
ρ_p = particle density
σ_y = horizontal dispersion coefficient, m
σ_Z = vertical dispersion coefficient, m
ϕ = angle
μ_a = air viscosity
μg = microgram
μ = micron, 10^{-6}m
χ = shape factor

INDEX

absorbents 70
activity factor 31
add-on equipment 41
adsorber 70
agricultural operations 67
air eddies 14,57
angstrom 7
area source 125
ash analysis 94
asphalt concrete 72
asphalt plant 74

baghouse 65
best available control technology
    (BACT) 41
beta gauge sampler 113
binders 53,71
boilers 71

capture velocity 61
"carpets" 55
cement 74-75
    manufacturing 74
centerline velocity 61
chemical binders 53
chemical oxidants 70
chemical stabilizer 54,71
chemical suppressants 48
Civil Engineering Fabrics 55
climatic data 31
coagulation 47
coal
    analysis 94
    handling 64
    pile 91
coalescence 47
combustion sources 71
confinement 54

construction 15
conveyance 13
correction factors 31
covers 55
CRSTER 96,125
crushing 30
cyclone 70

decay 69
    aerobic 69
    anaerobic 69
"default" 95
deposition velocity 126
diffusion coefficients 28
diffusion models 125
dispersion equations 124
dispersion models 95
drift 10
dry suppression 53
    electrostatic 53
    sonic 53
duct connections 63
duct velocity 61
dust
    capture 61
    deposition 125
    suppression 46

electrostatically charged fog
    atomization 47
electrostatic suppression 53
elevated monitors 113
emission controls
    asphalt plant 74
    assessment 40
    cement plant 74,119
    combustion 72
    construction 20,30

conveyance 20
cooling tower 92
evacuation 20
foundry 76
gaseous 5,27
grain handling 76
industrial 39,115
lime plant 77,79
mining 78,116,117
ore handling 80
power plant 89,91,93,102
road dirt 16-17,32,71,124
size reduction 30
smelting 103
stoker 72
storage piles 23
taconite handling 84
tillage 19
transfer 20
utility 89,91,93,98
woodworking 104
emission standards 41
encrusting 5,71
equivalent diameter 61
estimating control efficiencies 116, 122

fabrics 55
face velocity 61
field operations 67
filter 65
fogging 47,65
foundries 75
free stream velocity 8

grain 70,75
  dusts 70
gravel 71
gravitational settling 125
ground level concentrations 31,95, 113
ground level monitors 113
growing plants 69

health effects 39
high-volume air samplers 112

high-volume method 131
hoods 61
hood design 61
hood velocities 62
housekeeping 45
humidification 46,47

impact analysis 96
inhalable particles 1,39,112
inlet duct 64
isotachs 57

lime plants 75,78
limestone crushing 78
line sources 124
linings 55
loading 13
load-in stacker 22
lowest achievable emission rate (LAER) 41

measurements 107,115
mining 40,77-78
  coal 77
  copper 77
models 124
moisture content 31
monitoring 107,108

National Ambient Air Quality Standards (NAAQS) 1
nonferrous smelting 104

odors 5,70
offset policy 83
ore handling 78
oxidizing acids 70
oxidizing bases 70

particulate report 127
particle size 31
particle travel 10
phosphate 77
pile orientation 57
pollutant criteria 108
portland cement 75

power plants 87
precipitation 5
prescription burning 72
preventative procedures 41,45
Prevention of Significant Deterioration (PSD) 4
process criteria 108
process modification 41,43
process rates 31

quasi-stack sampling 109

radioactive nuclides 94
rain 5
reasonable available control technology (RACT) 41
respirable particles 1,39
roads 13,16-17,32,71,124
   silt content 17
roof monitor sampling 109

saltating particles 7
sector selector monitoring 112
settling 125
shapes 8
silt 17,25,31
site criteria 108
size reduction 15
size-specific particulate standards 1
sizing 13
soaking spray 45
soil porosity 28
sonic suppression 53
spraying 45-47,53,65,68,78
   chemicals 47
   coverage 47
   liquid density 53
   nozzles 45-46, 68
   pressure 68
stabilizing chemicals 53-54, 71
stone quarrying 77
storage
   duration 31
   pile orientation 25

   piles 14,25
suppression
   chemical 48
   electrostatic 53
   sonic 53
   wet 45, 47, 65
surface creep 7
surface roughness 126
suspension velocity 8

taconite ore 32
"terminal velocity" 7
terrain 96
"threshold friction velocity" 7
tillage operations 13,19
   agricultural 19
total suspended particulates (TSP) 1
toxic substances 102
tracer sampling 114
transport 5,13,20,62
   ducts 62
   velocity 62
turbulent air eddies 14,57

unloading 13
upwind-downwind sampling 110

VALLEY Model 95,195
vapors 5
vehicles 13
velocity profile 10
visibility problems 39

waste sites 14,27
wet scrubber 70
wet suppression 45,47,65
wind
   control 54
   data 95
   erosion 5,72
   matrix 102
   speed 57
windbreaks 56,124